I0829266

HIGH MASONRY

DAMS.

BY

JOHN B. McMASTER, C. E.

AUTHOR OF "BRIDGE AND TUNNEL CENTRES."

NEW YORK:
D. VAN NOSTRAND, PUBLISHER,
23 MURRAY AND 27 WARREN STREET.
1876.

PREFACE.

In the preparation of the following treatise three points have been constantly in view, to avoid as far as possible all purely theoretical discussion, to discover the most economical forms of profiles consistent with perfect strength, and to consider none that have not, after repeated practical application demonstrated their excellence, even under the severest tests. To treat the subject, however, in a logical way, it has been found best to begin with a theoretical determination of the strongest and at the same time least expensive form of profile, and afterwards to modify this to meet the requirements that arise in actual construction.

The theoretical type is, as I have attempted to show, that composed of a vertical face on the inner, and a concave surface on the outer side. Of this, there are, of course, an almost unlimited number of possible modifications. But when we impose the condition of *economy*, the number of really useful ones dwindle down to less than half a dozen. Those treated of in the present work number four. The first (illustrated in Fig. 9), is, beyond all doubt, the very best. It has indeed, been often urged against this type of profile, that it is difficult to determine with accuracy the equations of the logarithmic curves forming the bounding faces, as also to cut the facing stones to such a curve. As to the first objection, no equations can surely be simpler than those we have given, while the second is a difficulty most easily removed, not by argument, but by determination.

The three other types are also profiles of equal resistance, and are treated of so fully in the work as to call for no remark here. It will also be observed that I have touched very lightly on the sliding of dams on their foundation, or of any portion of them along a horizontal joint. This has been done, because, though I have examined as fully as possible the causes that have led to the destruction of dams of all style of profile and of all heights, both abroad and in this country, I have been able to find extremely few that may justly be said to have yielded by sliding. It has almost invariably been by revolving about an axis near the outer face, caused by taking too great a limit of vertical pressure, and thus throwing the line of resistance, when full, too far outward from the centre of thickness.

JOHN B. McMASTER.

New York, February, 1876.

HIGH MASONRY DAMS.

The subject proposed for consideration in the following work is that of the profile of masonry dams of such height, breadth and general dimensions as would be required for reservoir purposes, or for impounding the waters of rivers and large streams for mill or irrigation use. We would observe, however, at the outset, that as this matter has already been treated with such fullness by several writers, and especially by MM. Delocre and Sazilly—to whose excellent "memoirs" we are greatly indebted—we can hope to add little that is really new, but shall endeavor, by drawing from many sources, to supply our own deficiency, to diminish the errors of others, and thus obtain results very much more accurate than could be derived if we relied solely on ourselves.

Before, however, we take up the consideration of the matter of the form of profile that shall combine the greatest strength with the least amount of material, there are a number of important points to be considered somewhat in detail. Thus, it is necessary, in the first place, that we should know the forces to which dams are subjected, their kind, whether constant or variable, the methods of determining their direction and calculating their intensity, and the effects they are likely to produce, and these matters being known, we may pass to the consideration of the conditions of stability, first when the dam has only its own weight to support, and, secondly, when it has to withstand both its own weight and the pressure of the water. We may then deduce a theoretical profile of equal resistance, and, finally, adopt one so modified by the requirements of practice and suggestion of experience, that it shall serve as a *profile type*, fulfilling to the utmost the requirements of great strength and stability, beauty of outline and economy of material.

Now, it becomes evident, after a moment's consideration, that there are but two forces that may at any time be regarded as acting with vigor on a dam, and these are, the weight of the masonry, cement and other material composing the structure, and the pressure or thrust of the water whose flow it checks. The first becomes, to all intents and purposes, a constant quantity as soon as the dam is finished, and continues so for ever after, acting vertically downwards through the centre of gravity of the mass. But, on the other hand, the latter force is one of great variability. For, as its intensity at any moment depends on the depth or head of water behind the dam, increasing as the water deepens and decreasing as the water falls, and the head of water, especially in reservoirs used for mill or irrigation purposes, being subject to frequent rise and fall, it follows that this thrust must be considered as a variable quantity and treated accordingly. It is, moreover, to be observed, that this thrust acts horizontally, and unlike the weight, is not distributed uniformly over the entire face of the dam, being almost, if not quite, zero at the point where the water cuts the masonry, and growing greater and greater as we descend towards the foot of the dam. The weight, it is true, also increases as we go from the top to the bottom, yet, if we suppose the dam to be at any point ten feet thick, the pressure on any horizontal section taken at that point will be everywhere the same, and this is by no means the case if we take an area ten feet square on the water face of the dam, and against which the fluid presses.

In order that the dam may not yield under the first force, and be thrown down by the greatness of its own weight, it is necessary, should the structure be of such height, or the material of such heaviness, that the pressure per unit of surface at any horizontal section is in excess of the "limit of pressure" for masonry, that the surface of the section be increased so that the pressure being distributed over a more extended area the load at each unit of surface shall be less. The second force, or thrust of the water, is resisted at any point by the weight of the masonry above that point, and by the friction of the stones, which is of course dependent on the weight. Some resistance is indeed afforded by the bonding power of the hydraulic mortar used in setting the stones, but this is so small that precautions of safety require that it shall in all calculations be disregarded entirely.

But these two forces, the weight acting vertically downwards and the thrust of the water acting horizontally, counteract each other to a certain extent, and give rise to a third power or resultant, the position of which, as regards the base, will determine the stability of the dam. To illustrate, let A B C D (Fig. 1) represent the profile of a dam composed of horizontal courses of masonry bedded on each other, and K the centre of gravi-

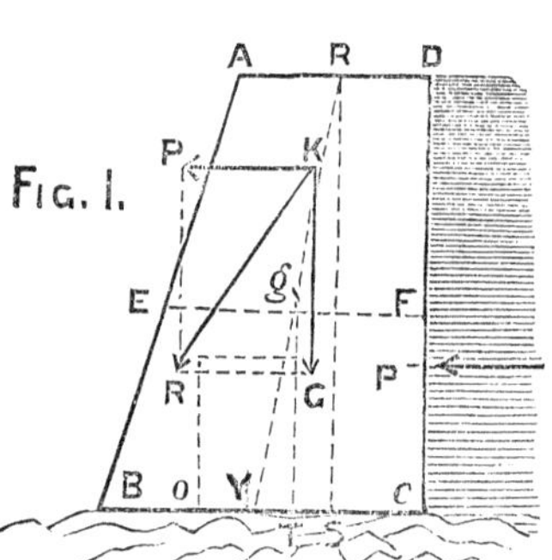

ty of the mass, lying above the line E F. Represent by K G the direction and intensity of the weight of A F, and by K P the direction and intensity of the thrust of the water from D to F. Then, constructing in the usual way the parallelogram P K G R, we shall have for the resultant of K P and K G, the line K R. Now, supposing the dam to be perfectly secure as to its weight, the force P of the water can demolish the wall only, when, exceeding the weight and friction K G, it shoves the mass A F along the joint E F, or causes it to rotate about an axis through E. Which of these motions, the slipping or rotating, shall take place depends entirely on the magnitude and direction of the resultant K R. If the pressure of the water is so large compared with the weight that the angle R K G, which the resultant makes with the vertical, is larger than the angle of friction (32° for masonry on masonry), the mass A F will then *slide* along the line E F; while if the position of the resultant is such that it passes without the base B C, then rotation will take place about the axis of E. Of these two motions, the latter is in practice the most likely to occur, inasmuch as in nine cases out of ten when rotation does take place it does so about some point as E′, nearer the resultant than E, because the pressure concentrated at E, breaks off the stone, and thus throws the axis of rotation nearer the resultant.

The condition of stability then, in dams that do not transmit laterally to the sides of the valley, the pressure they sustain (and this is the case in all large dams) is, that they must resist this pressure at every point by their own weight. If the material employed were of considerable resisting power, as well as the soil of the foundation, and if there were between them an unlimited degree of adhesion, the only condition of stability to be fulfilled would be, as we have just seen, to give the wall such a profile that the resultant of the thrust of the water and the weight of the dam shall pass within the polygon of the base. But this condition is not found sufficient in practice ; the material and the soil of the foundation will, in fact, support only a limited pressure (depending on their nature), and they have not between them an unlimited degree of adhesion. Hence, the two following indispensable conditions :

1° The several courses of masonry in the wall must be incapable of slipping the one over the other, and the wall incapable of sliding on its base.

2° In no point of the structure may the material employed, or the soil of the foundation be required to bear too great a pressure. To begin with the first condition.

STABILITY AS TO SLIPPING.

We shall take up first the condition of stability as to the slipping of the various courses of masonry, and then pass to that of the entire dam. The first thing to be now determined, is the horizontal thrust of the water. Suppose A B C D (Fig. 2) to represent the face of a dam pressed by water, and let h=A J denote the height; a=J C the projection of the slope of the dam on the horizontal plane; and, finally, let l=A B denote the length of the dam, and b=A G is breadth across the top. Then will the *vertical* pressure of the water on the face A B C D be expressed by

$$a\,l\frac{h}{2}\,y=\tfrac{1}{2}\,a\,l\,h\,y \quad . \quad . \quad . \quad 1$$

and the horizontal *thrust* by the expression

$$l\,h\frac{h}{2}\,y=\tfrac{1}{2}\,h^2\,l\,y \quad . \quad . \quad . \quad 2$$

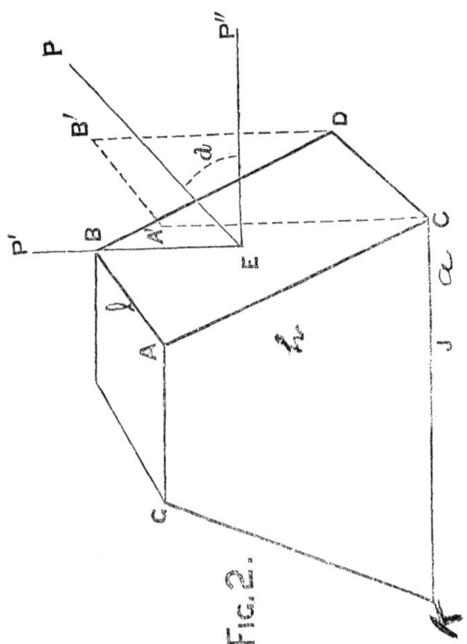

Fig. 2.

in each of which y denotes the density of the water. These equations are obtained as follows :

Let E P, in Fig. 2, represent the normal pressure of the water on the surface A C, which we will call F, and resolve it into two components, one vertical EP', and one horizontal EP'', and call them respectively P' and P''. Then expressing the angle PEP'' made by the horizontal component P'' and the normal EP, by a we shall have from the triangle EPP''

$$\frac{PP''}{EP} = \sin PEP'' \text{ or } \sin a.$$

But $PP'' = EP' = P'$, hence

$$\left.\begin{aligned} &\frac{P'}{P} = \sin a \quad \text{or } P' = P \sin a. \\ &\text{In the same way we find} \\ &\frac{P''}{P} = \cos a \quad \text{or } P'' = P \cos a. \end{aligned}\right\} \quad . \; . \; 3$$

Now, let a projection $A'B'CD$, of the surface $ABCD$, be made on a plane at right angles to P'', and call the area of the projected surface F'. Then will $F' = F \cos ACA'$, or since the angle of inclination ACA' of the surface to its projection is equal to the angle $PEP'' = a$, between the normal to AC, and the perpendicular to $A'C$, we shall have $F' = F \cos a$ or $\cos a = \frac{F'}{F}$. But $\cos a$ is by equation 3 equal to $\frac{P''}{P}$, and therefore,

$$\frac{P''}{P}=\frac{F'}{F} \quad \text{or } P''= P\frac{F'}{F} \quad . \quad . \quad 4.$$

From the principles of mechanics, we know that the pressure P of water on any given area is the product of the area, the height h of the water, and its density y, so that in the present instance F being the area of the surface A B C D, we shall have for the value of P the expression $P=F\,h\,y$, and this substituted in equation 4 gives

$$P''=F\,h\,y\frac{F'}{F} \quad \text{or } F'\,h\,y \quad . \quad . \quad . \quad 5.$$

Therefore is the pressure with which water presses against a surface in a given direction equal to the weight of a column of water, which has for its base the projection of the surface pressed, and for height the depth of the centre of gravity of the surface below the top of the water. We see, moreover, from the above, that since the projection at right angles to the vertical is the horizontal, and the projection at right angles to the horizontal is the vertical projection, the vertical component of the pressure of water against a surface may be found if the *horizontal projection*, or its *trace*, be considered as the surface pressed, and, on the other hand, the horizontal component may be found if the *vertical projection* of the surface, or its *trace*, be considered as the surface pressed.

Applying these two principles to the case of Fig. 2, and replacing F' in equation 5, by its value $l\,h$, we shall have for the horizontal thrust of the water on the face A B C D of the dam the equation $P''=\frac{1}{2}\,h^2\,l\,y$, and in the same way the vertical component will be found to be equal to $P'=\frac{1}{2}\,a\,h\,l\,y$. Now, b being the breadth of the dam, and a' the projection of the slope G K, and y' the density of the masonry composing the dam, it is evident that the area of K C E G will be $\left(b+\frac{a+a'}{2}\right)\,h$; the cubic contents $\left(b+\frac{a+a'}{2}\right)hl$, and the *weight* $\left(b+\frac{a+a'}{2}\right)$ $h\,l\,y'$. The whole vertical pressure on the base will therefore be equal to this weight *plus* the vertical pressure of the water, or

$$\tfrac{1}{2} a l h y + \left(b + \frac{a+a'}{2}\right) h l y'$$
$$= \left\{ \tfrac{1}{2} a y + \left(b + \frac{a+a'}{2}\right) y' \right\} h l \quad . \; 6.$$

We have seen, however, that the force which tends to counteract the push of the water, and on which the stability as to slipping must therefore depend, is equal to this weight of the dam increased by the *friction* of the stones. Denoting this co-efficient of friction by f, we shall then have for the force to push the dam forward the expression

$$\left\{ \tfrac{1}{2} a y + \left(b + \frac{a+a'}{2}\right) y' \right\} f h l$$

and in the case where the horizontal thrust of the water is to effect the displacement

$$\tfrac{1}{2} h^2 l y + \left\{ \tfrac{1}{2} a y + \left(b + \frac{a+a'}{2}\right) y' \right\} f h l$$

or dividing each member through by $\tfrac{1}{2} h l y$, we shall have

$$h=f\left\{a+\left(b+\frac{a+a'}{2}\right)\frac{y'}{y}\right\} \quad . \; 7.$$

In order therefore that the dam may not be pushed away by the water, we must have one of the two following conditions fulfilled; either

$$\left.\begin{array}{l} h<f\left\{a+(2\,b+a+a')\dfrac{y'}{y}\right\} \\ \text{or} \quad b>\frac{1}{2}\left\{\left(\dfrac{h}{f}-a\right)\dfrac{y'}{y}-(a+a')\right\} \end{array}\right\} \quad 8.$$

For safety, we may further assume that the base of the dam is quite permeable, in which case there is (on the principle that a pressure in one direction produces an equal pressure in the opposite direction) a pressure from *below* upwards equal to $(2\,b+a+a')\,l\,h\,y$, equal the weight of the dam, and as this is, of course, to be subtracted from the above, we have finally,

$$h<f\left\{(2\,b+a+a')\left(\frac{y'}{y}-1\right)-a'\right. \quad 9.$$

These equations are applicable not only to the sliding of the entire dam on its foundation, but also to any particular layer of stone at any point in the dam. The value of the co-efficient of friction *f* will of course be very different in cases where we consider the stability of different parts of the wall, from that in cases where we consider the dam to slide on an earthty foundation. In the former case, it is that of masonry on masonry, in the latter, that of masonry on earth, and in general clay. In fact, it may be restricted almost solely to clay, because in a sandy, porous or yielding soil, it is better, on principles of economy, not to build a dam, but a dyke. For masonry on masonry, or, indeed, bricks on bricks, we may with safety take the co-efficient of friction as equal to .67 ; for masonry on dry clay .51; but for masonry on wetted clay the co-efficient falls to .33.

A few examples may, perhaps, serve to illustrate the above remarks. We shall confine ourselves first to the case of rotation about one of the joints, as that is really the most likely one to arise in practice :

Let Fig. 1a represent the profile of a

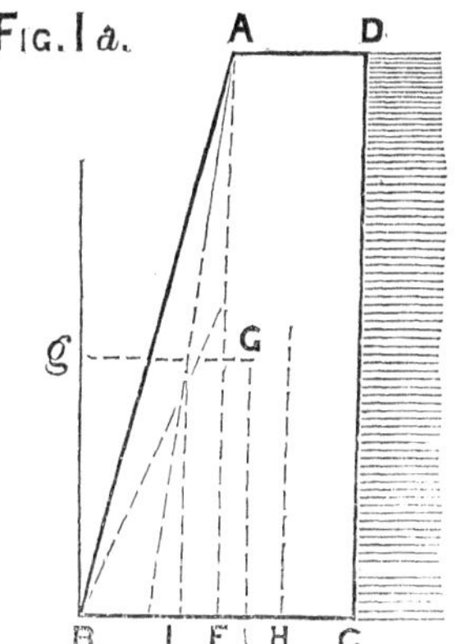

dam, constructed say of brickwork weighing 112 pounds per cubic foot. Let the thickness on top be 10 feet, and that at the base 20 feet, required to find the perpendicular height, the dam must have in order that, when the water stands at the brim, the wall shall be just on the point of turning about the point B under the pressure of the water. Denote by h the height of the dam, or the quantity we are in search of, = C D. Now, by equation 2, the thrust of the water on one lineal foot of surface is $\frac{h^2}{2} \times 62.5$ lbs., and the *moment* of this thrust is $\frac{h^2}{2} \times 62.5$ lbs. $\times \frac{h}{3}$ or $\frac{h^3}{6} \times 62.5$ lbs. The pressure of one foot of the dam, or what is the same thing, its weight is $\frac{AD+BC}{2} h \times$ 112 lbs., or $\frac{10+20}{2} h \times 112$ lbs. $= 1680\ h$ lbs., and the *moment* of this pressure with reference to the point B is $1680\ h \times BE$. Before we can obtain this moment, then, we must find the value of B E, and this is found as follows :

It is evident from a moment's inspection of Fig. 1*a*, that the area of

$$ABCD \times Gg = \text{area } ABCF \times BH + \text{area of } ABF \times IB, \text{ or}$$

denoting A D by a; B C by b; D C by c; and Gg by d, we have since $BH = \frac{2b-a}{2}$, and $IB = \frac{2(b-a)}{3}$.

$$c\frac{b+a}{2}d = ac \times \frac{2b-a}{2} + \frac{c(b-a)}{2} \times 2(b-a)$$

dividing by c

$$\frac{b+a}{2}\,d = a \times \frac{2\,b-a}{2} + \frac{(b-a)^2}{3}$$

$$\text{B E} = d = \frac{2\,b}{3} - \frac{a^2}{3\,(a+b)}$$

Substituting for the above quantities their values, we have :

$$d = \tfrac{40}{3} - \tfrac{100}{90} = \tfrac{110}{9}.$$

The moment of the dam therefore is $1680\,h \times \frac{110}{9}$.

$$\therefore\ \frac{d^3}{6} \times 62.5 \text{ lbs.} = 1680\,h \times \tfrac{110}{9}.$$

$$\frac{62.5\,h^3}{6} = \frac{184800\,h}{9}$$

$$h^2 - \sqrt{197.12}$$

$$h = 44.3982.$$

Again, preserving the same dimensions, let it be required to find the "modulus of stability" of a masonry dam of the profile, shown in Fig. 1, the stone weighing 200 pounds per cubic foot. Draw from the middle of the top A D to the middle of the base B C the line R V, and take its length as 45 feet, and the depth of the water behind the dam, 44 feet.

Now, by geometrical principles, which it is not worth while to repeat here, we have :

$$V\,g = \tfrac{1}{3}\,R\,V\,\frac{B\,V + A\,D}{B\,V + A\,R} \text{ or } V\,g = \frac{45}{3} \times \frac{10+10}{15} = \frac{245}{15}$$

g being the centre of gravity of the wall. Again, in the two similar triangles R V S and g V T, we have :

$$R\,V : V\,S :: V\,g : V\,T.$$

The value of V g we have just found. V S is evidently equal to V C—S C, or 10—5=5. In the triangle. R V S, we also have $R\,S^2 = R\,V^2 - V\,S^2$, or $R\,S^2 = (45)^2 - (5)^2$; hence R S=44.38. Substituting these values in the above proportion, we shall have :

$$45 : 5 :: \frac{245}{15} : V\,T \quad \therefore \quad V\,T = 1\tfrac{2}{3}$$

The weight or pressure of the wall acting through the centre of gravity g of the dam is, as we have already seen,

$$\frac{20+10}{2} \times 1 \times 44.38 \times 200 = 133140 \text{ lbs.}$$

and that of the water $44 \times 1 \times \frac{44}{2} \times 62.5 = 60500$ lbs. If now we denote by P the "centre of pressure" of the water, that is to say, that point where a single pressure will counterbalance the thrust of the water against the entire face D C of the dam, then $P = C\,P = \frac{44}{3} = 14.6$ feet. The quantity we are in search of, the modulus of stability of the wall is the ratio of T B to T O. The value of T B we have already, and may obtain that of T O from the proportion that the pressure of the dam is to the height of the centre of pressure (P) of the water above the base of the dam as the pressure of the water is to the entire pressure of the water acting on its centre of pressure P. Thus:

$$133140 : 14.6 :: 60500 : x$$
$$x = 6.6 = T\,V.$$

Dividing this last found quantity by T B, we have:

$$\frac{T\,V}{T\,B} = \frac{6.6}{11\frac{2}{3}} = .537 = \text{modulus of stability.}$$

In a well built structure, this quantity

should never be *less* than .5, hence, as in the present case, the modulus is somewhat above this value, we are justified in regarding the dam as a perfectly stable structure, when the water is not over 44 feet in depth.

In these considerations, we have taken no account of the resistance offered by the adhesion of the mortar. Should this be taken into account—and it is always best that it should not—then equation 9 will require to be modified somewhat as follows : Let H equal the distance of the centre of gravity of a layer of stones below the top of the dam. The shove of the water tending to throw down this portion of the dam is, as we have just seen, $\frac{\delta H^2}{2}$, in which expression δ is merely a short notation for ly. The forces resisting this shove are the friction of the two layers sliding on each other, and the adhesion of the masonry. The first is proportional to the weight of the masonry above the stratum in question, and the second or adhesion of the masonry is proportional to the thickness of the dam at this point. Representing as before the co-efficient of friction by f, by c the cohesion of the mortar per unit of surface, by s the area of the upper surface of the course next below, and by b the thickness of the dam at this section, we shall have for the resistance R to sliding :

$$R = s\,\delta' f + c\,b$$

and therefore, in order to insure stability, we must have :

$$s\,\delta' f + c\,b > \frac{\delta\,H^2}{2}$$

or clearing of fractions, and then dividing by $\delta\,H^2$,

$$2\frac{(s\,\delta' f + c\,b)}{\delta\,H^2} > 1 \quad . \quad . \quad . \quad 10.$$

Neglecting the adhesive power of the mortar, the above becomes :

$$\frac{2\,s\,\delta' f}{\delta\,H^2} = 1 \quad \text{or } f = \frac{\frac{\delta\,H^2}{2}}{s\,\delta'}$$

The second case of slipping, or that of the dam on its foundation will rarely, if ever arise, when the dam is founded on a rock, for in that case the value of the co-efficient of friction will be the same for the horizontal section of the foundation as for any section of the masonry. It is, however, very likely to arise whenever circumstances will not enable us to lay the foundation on bed rock. In such cases the soil will almost always be of an argillaceous nature, for, should it prove to be of a gravelly, sandy or very permeable character, the employment of some common form of dyke will be much preferable to the construction of a dam. We may, therefore, reasonably assume that in all cases where the foundation course does not rest on a rock surface, it will be laid on argillaceous soil, and as this will readily give, under the action of water, a slippery slimy surface, we must assume a co-efficient of friction very much less than that used for masonry on masonry. With this point kept clearly in view, the conditions of stability will be given by the above equations. Yet there are one or two other considerations that must not be overlooked. Thus, as the stability will depend in large measure on the lateral resistance of the soil, it is not sufficient to be sure that this resistance is large enough to prevent the sliding of the wall, but is also necessary to be assured that at any point of the front of the foundation wall, the normal pressure does not exceed the limit R′ of which the soil or the wall is susceptible. Again, in order to prevent any slipping likely to arise from the lateral compression of the earth, it is not necessary to interpose any packing between the face of the wall and that of the ditch, and, finally, that in all cases it never comes amiss to "step" the rock or the earth on which the foundation course rests, a matter to be considered more in detail hereafter.

SECOND CONDITION OF STABILITY.

To return now to the second condition of stability, namely, that in no point of the structure may the material employed, or the soil of the foundation, be required to bear too great a pressure. For this purpose let A B C D (Fig. 3) repre-

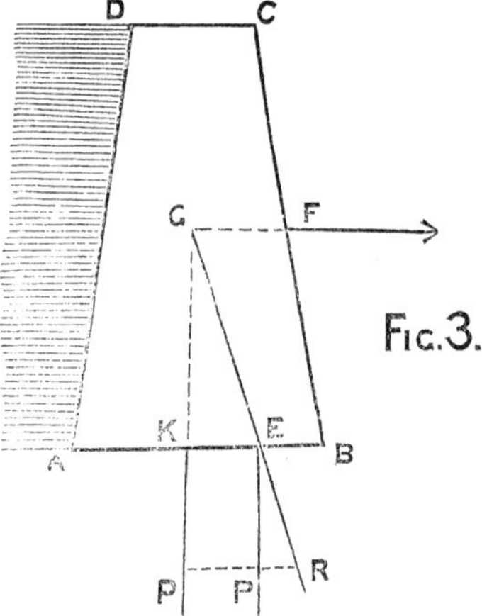

FIG. 3.

sent the profile of a dam. Then from the principles we have already established, it follows that any section of this, equal in length to a lineal unit, may be considered as subject to the action of two forces, which are, respectively, the vertical component P of the *resultant* of the weight of the structure above that unit, and the horizontal pressure or *thrust* of the water, and the horizontal component F of the thrust of the water. In the section A B C D, these two forces act through the centre of gravity G, and produce a resultant of their own which cuts the A B at E. This latter resultant R may therefore be regarded as applied directly to the point E, and resolved into two components, one vertical and equal to the force P, and one horizontal and equal to the force F. The horizontal force tends to slide the wall along the base A B. This we have considered. The vertical spreads itself over the base from the extremity B, which is nearest the point of application of the resultant, according to the well known decreasing law. Now, in all works on mechanics, we have given a formula which applies to a homogenous rectangle, pressed by a force acting upon one of the symmetrical axis, and this is :

$$p = \frac{N}{\Omega}(1+3n) \quad . \quad . \quad . \quad (\propto)$$

and

$$p' = \frac{N}{\Omega}\left(\frac{4}{3(1-n)}\right) \quad . \quad . \quad (\beta)$$

Where N is the entire load or pressure, and Ω the entire area of the surface pressed. In the case we are considering, the quantity N in equations $\propto$ and β, is, of course, represented by P the vertical component. Ω, by l, if by this letter we designate the breadth of the base A B, and if we denote the distance E B by u, then will the quantity n in equations $\propto$ and β be represented by $\frac{l-2u}{l}$.

Substituting these quantities, we shall have :

$$p = \frac{P}{l}\left(1 + \frac{3l-6u}{l}\right) = \frac{P}{l}\left(\frac{l+3l-6u}{l}\right)$$

$$= 2\left(2 - \frac{3u}{l}\right)\frac{P}{l} \quad . \ . \ . \ . \ . \quad 11.$$

and

$$p = \frac{P}{l} \times \frac{4}{3\left(1 - \frac{l-2u}{l}\right)} = \frac{P}{l} \times \frac{4l}{6u}$$

$$= \frac{2P}{3u} \quad . \ . \quad 12.$$

Equation $\propto$ is applicable in all cases where $n < \frac{1}{3}$, and therefore equation 11 is applicable when $\frac{l-2u}{l}<\frac{1}{3}$; that is when $u>\frac{1}{3}l$.

Equation β is applicable to all cases when $n>\frac{1}{3}$, and consequently equation 12 to all cases when $\frac{l-2u}{l}>\frac{1}{3}$, or, what is the same thing when $u<\frac{1}{3}l$. We have seen that the condition of stability requires that some limit, R′, should be placed on the pressure each superficial unit is expected to bear. The pressure at the point B, must therefore be less or never greater than R′, and we shall have according as u is greater or less than $\frac{1}{3}l$,

$$2\left(2-\frac{3u}{l}\right)\frac{P}{l}=\text{or}<R' \quad . \quad 13.$$

and

$$\frac{2}{3}\frac{P}{u}=\text{or}<R' \quad . \quad . \quad . \quad . \quad . \quad 14.$$

And this condition is to be fulfilled for each section made in the profile, neglecting the force of cohesion of the mortar which is unfavorable to resistance.

These expressions are susceptible of yet further modification, if we introduce into the calculation the maximum height λ that may be given to a wall with vertical faces, so that the pressure upon the base shall not exceed the limit R′ of safety. Indeed, if we represent the density of the masonry, or the weight per cubic yard by δ, we shall have $R' = \delta' \lambda$, and the above equation become:

$$2\left(2 - \frac{3u}{l}\right)\frac{P}{l} = \text{or} < \delta' \lambda = 2\left(2 - \frac{3u}{l}\right)$$

$$\frac{P}{\delta' l} = \text{or} < \lambda \quad . \quad . \quad 15.$$

and

$$\frac{2}{3}\frac{P}{u} = \text{or} < \delta' \lambda = \tfrac{2}{3}\frac{P}{u\,\delta'} = \text{or} < \lambda \quad . \quad . \quad 16.$$

The conditions expressed in these equations would be quite sufficient if the water was always up to the top of the dam, but as this is by no means always the case, the wall must be capable, even when the dam is quite empty, of supporting its own weight without being subject at *any* point to a pressure per unit of surface exceeding the limit $\delta' \lambda$.

In this case the resultant of all the

forces acting on the wall is reduced to the weight P′, and denoting by K A, the distance from the resultant passing through the centre of gravity of Fig. (3) to the nearest extremity A of the base, by u, the pressure at A, will be given according to circumstances by equations 11 or 12, and the *stability* of the wall will require that one of the relations expressed in equations 15 or 16 be satisfied when P′ is substituted for P.

The next step, therefore, is to determine the proper

PROFILE FOR A DAM HAVING ONLY ITS OWN WEIGHT TO CARRY.

In order to study under all conditions, the question we are now about to consider, it is perhaps well to inquire, in the first place, what form it is most *convenient* to give a dam having only its own weight to carry, in order that each point of the masonry shall not be subjected to a pressure larger than the limit of safety, and then to determine the alterations which economy require to be made in this *assumed* profile. It is evident, to begin with, that when the height of the dam is such that it does not go over the limit λ (*i. e.* the greatest height we can give to a vertical wall, without the pressure on the base becoming larger than R′, we shall be quite justified in giving the dam vertical facings, and that, in such case, the load for each unit of surface at the lower part will be somewhat less than $\delta'\lambda$, or at least, never greater. Again, we know that whenever the pressure on a horizontal surface of masonry is larger than the limit of safety, we may correct this, by enlarging the area of the surface pressed, and so lessen the load on each superficial unit. And these are the two fundamental principles of dam construction, and may be summed up in brief as follows: If we are constructing a dam of a height equal to or less than λ, *and having only its own weight to support*, it is a safe practice to give it vertical facings from top to bottom. If, however, we are constructing a dam of a height *greater* than λ, *yet having only its own weight to support,* we must make the faces vertical for a distance from the top equal to λ, and *from* this point to the *base* slope them *outward.*

A dam constructed on this latter principle would give a profile similar to that in Fig. 4. From the summit A B to the section C D, the pressure per superficial

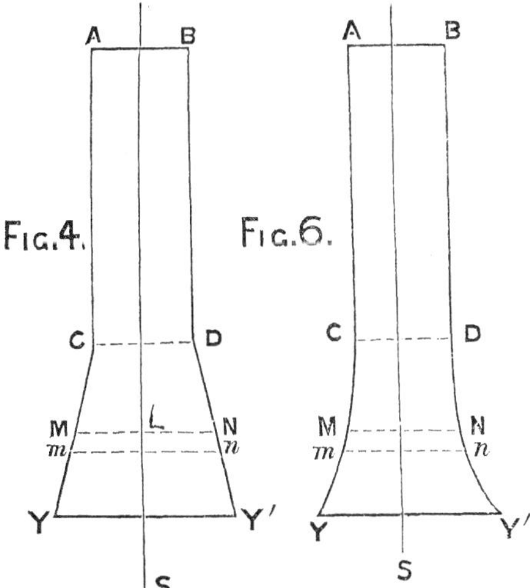

unit is nowhere greater than $\delta'\lambda$, and

therefore from A to C the face is vertical, but *below* C D, the load exceeds the limit and increasing at each section to the base, and hence from C to Y the face is sloping. And just here we are met by the great question in dam construction that of *profile*. Should the bulging portion C Y Y′D, be bounded by right lines as in Fig. 4, should it be *stepped*, should it be curved, and if so, should the bounding curves be logarithmic curves, simple or compound? these are questions we propose to consider.

It is an easy matter to determine the force to be given to the facing, so that the condition that the load per nnit of horizontal surface shall never go over the limit $\delta'\lambda$, shall be satisfied. To do this, we may choose arbitrarily one face and then determine the other, but if we desire to use the minimum of material consistent with perfect safety, then the wall must be symmetrical as to its axis. In such a case as that illustrated in Fig. 4—that of a high masonry dam, whose height is greater than λ—the slopes

D N Y′ and C M Y, ought to satisfy the requirement that, if in any section, as M N, the load per surface unit is equal to any given quantity, the pressure will be the same for any other section as $m' n'$, infinitely near to it. This will be fulfilled, if the increase given to the base is proportional to the increase of pressure, or as the profile is to be made symmetrical to the axis O S, if the increase of the half surface L N or L M is proportional to the increase of load on that half surface. If we denote by P the pressure on L N, arising from the weight of the structure above, and a the surface of this section, then, it is evident, the above condition will be expressed by

$$d\,P = K.\,d\,a. \quad . \quad . \quad . \quad 17.$$

In which K is a constant quantity, and denotes the limit of pressure on the unit of surface or $\delta'\lambda$. Again, by b, denote the dimensions of the dam in the direction perpendicular to the section we are concerned with, and by x the length of the half section L N, or, to express it mathematically, the abscissa of the curve or line sought (*i. e.* D N Y′), and finally, by y, the distance of M N from a horizontal line taken as the axis of x. Then the surface a will equal to $b x$, and consequently an increase of surface as $d\,a$ in equation 17, will be expressed by

$$d\,a = d\,b\,x$$

and moreover

$$d\,\mathrm{P} = \delta' \, b\,x\,d\,y$$

These values substituted in equation 17 give for the differential equation of the curve,

$$\delta' \, b\,x.\,d\,y = \mathrm{K}.\,b.\,d\,x \quad . \quad . \quad 18.$$

whence

$$d\,y = \frac{\mathrm{K}}{\delta'} \, \frac{d\,x}{x}$$

But K equals the limit of pressure per unit or $\delta' \lambda$, and this value replaced for K, we shall have

$$d\,y = \frac{\delta' \lambda}{\delta'} \, . \, \frac{d\,x}{x} \text{ or } d\,y = \lambda \, \frac{d\,x}{x}$$

Integrating this between the proper limits, we shall have

$$y - y_0 = \lambda \log \left(\frac{x}{x_0} \right) \quad . \quad . \quad . \quad 19.$$

Now, from this equation we see that, the curve being referred to rectangular axes, one of the co-ordinates is equal to the logarithm of the other, and, hence, the curve must be a *logarithmic curve*. Here then we have one property of the curve D N Y. To find in the next place the origin of its co-ordinates, we may make in the foregoing equations $x_0=\lambda$, in which case we shall have :

$$1=\frac{d\,y_0}{d\,x_0} \text{ and } y_0=0 \quad . \quad 20.$$

From this last relation it is quite apparent that the origin of co-ordinates is to be taken at a point where the value of x is equal to that of λ, and in this point the tangent to the curve makes an angle of 45° with the axis of x. Returning now to equation 19, let us replace y_0 and x_0 by their respective values, given in equation 20, when we shall have:

$$y=\lambda \log. \frac{x}{\lambda}$$

or passing from the system of Napier to the common system of logarithms,

$$y = 2.302658509\,\lambda \log. \frac{x^{*}}{\lambda} \quad . \quad . \quad 21.$$

This curve, when constructed, will give the form of the facing of a wall of indefinite height for which the pressure per unit of surface equals the limit of pressure K. It is not to be forgotten in making use of equation 21, that the direction in which y's are usually estimated has been reversed ; in other words, y when positive is to be estimated downwards, and when negative upwards, or in the direction of L O. Fig. 5 represents this curve constructed, by assuming the pressure limit or K as 132,000 lbs., and the density of the masonry as double that of water.

In such a profile, as Fig. 4 has, the sloping faces below C D being bounded by right lines, we may obtain the necessary breadth of the base Y Y′, as soon as we have determined the height and

---

* We may also pass from the Naperian to the common system, by multiplying the Naperian logarithm by the modulus of the common system, which is 0.434294. Its logarithm is 9.637784.

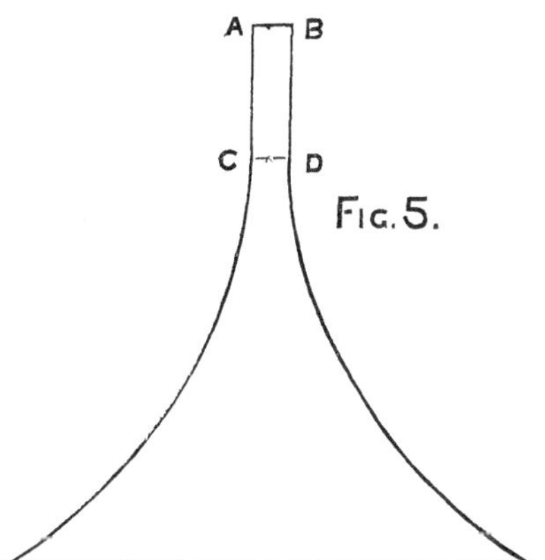

Fig. 5.

the breadth at top. Denote by b the breadth at top A B ; by h the distance A C$=\lambda$, and by h' the distance from C to the base Y Y′; by δ' the density of the masonry, and by x the quantity we are seeking for, or the base YY′. Then we shall have :

$$\left\{ h\times b+h'\left(\frac{b+x}{2}\right)\right\}\frac{\delta'}{x}=\delta'\times h \quad . \quad 22.$$

The quantity h in this equation, which is merely another expression for the quantity λ, has been determined by a number of investigators, but the most reliable results are those obtained by the French engineers,* who, in the construction of their great masonry dams, such as Furens, have taken the limit of pressure K at 60,000 kilogrammes, or about 132,000 lbs. per square metre, and K being equal to $\delta'\lambda$, and δ' being equal to 2,000 kilogrommes, λ becomes equal to 30 metres. As we shall hereafter see, however, the limit of pressure varies for the outer and inner face of the dam.

If, again, the profile adopted be such as is illustrated in Fig. 3, that is to say, if the faces of the dam slope continuously from the top to the bottom, then the thickness or breadth of the base will evidently be obtained by dividing the product of the height of the wall and its thickness on top by the difference between 2 λ and the height. For $\delta' \lambda$ or the limit of pressure is equal to the area of the profile, multiplied by the density of the masonry divided by the thickness of the base. In the figure, the area is

* MM. Delocre, Sazilly and De Graeff.

plainly equal to half the sum of the two parallel sides by the altitude, and denoting this latter by H, we shall, therefore, have:

$$\delta'\lambda = \text{H}\left(\frac{b+x}{2}\right)\frac{\delta'}{x} \quad \text{or } x = \frac{\text{H}\,b}{2\,\lambda - \text{H}} \quad . \quad 23.$$

The conditions which govern the construction of such a dam, and the height to which it is safe to build it, become from this equation quite apparent, should we make H$= 2\,\lambda$, then x would equal $\frac{\text{H}\,b}{0}$, and the base of the wall would spread out to infinity. Should we, upon the other hand, make H greater than $2\,\lambda$, then λ would become negative, and hence it follows that the greatest height we can give to a masonry dam with straight sides equally inclined from the summit and not go over the limit of resistance for masonry, is equal to twice that of a wall with vertical sides. Yet, within this limit, such a profile for a masonry dam of any height, occasions a gross waste of material. This becomes strikingly apparent, if we compare the breadth of base of a dam constructed with inclined faces from top to bottom, with that of a dam of the same height, but having a profile such as that of Fig. 4. Suppose each dam to be 30 metres high and 5 metres thick on top; required the thickness at the base. For the first case, using equation 23, we have :

$$x = \frac{30 \times 5}{60 - 30} = 5 \text{ metres.}$$

For the second form of profile, we use equation 22, and have, since the quantity h equals λ, the same value, or $x = 5$ metres.

If we raise the dam by 10 metres, then equation 23

$$x = \frac{40 \times 5}{60 - 40} = 10 \text{ metres.}$$

and by equation 23

$$\left\{ 30 \times 5 + 10 \left(\frac{5 + x}{2} \right) \right\} \frac{2,000}{x} = 60,000$$

or since

$$x = \frac{b\, \delta' \, (2\, h + h')}{\delta' \, (2\, h - h')} = x = 7 \text{ metres.}$$

If, once more, we add ten metres to the height, then equation 23

$$x = 25 \text{ metres.}$$

and eq. 22 $\quad x = 10$ metres.

The saving thus affected when the dams are of great height becomes simply enormous. The difference, however, between the profile when the dam below C D (Fig. 4) is bounded by right lines, and when bounded by logarithmic curves, such as shown in Fig. 6, is not so marked as in the cases just considered, yet is considerable. To take but one case in illustration, a dam of a profile such as Fig. 6 illustrates, with the faces below C D bounded by curves, would require (equation 21) a breadth of base equal to 9.739 metres, the height and thickness at top being as before, 50 and 5 metres respectively, while, as we have just seen, if the faces below C D were right lines, the base would be 10 metres.

Such, in brief, is the relative merit of these three forms of profile, for a dam having nearly its own weight to support.

In practice, however, such a dam can, of course, never exist, and it thus becomes necessary to take into consideration the second condition, or that of a dam supporting a charge of water.

PROFILE FOR A DAM RESISTING THE PRESSURE OF WATER.

And here, again, we are to throw aside, at first, all practical considerations, and determine a theoretical profile of equal resistance, one in every part of which the pressure shall not be greater than the limit R′. For this purpose we return to the two equations, deduced some time back, which express the conditions of stability for a dam resisting the thrust of water, and neglecting the signs > and < and the values corresponding to them, take only those corresponding to the sign of =. We then have the two following equations :

$$2\left(2-\frac{3u}{l}\right)\frac{P}{\delta' l}=\lambda \quad . \quad . \quad 24.$$

and

$$\tfrac{2}{3}\,\frac{P}{u\delta'}=\lambda \quad . \quad . \quad . \quad . \quad . \quad . \quad . \quad 25.$$

If we now replace the quantities u, l and P, by their respective values, expressed in functions of the height of the dam, we may readily deduce two equations which, on examination, will show two things.

1°. That the profile offering the least thickness, consistent with the conditions of stability, is one in which the side turned towards the water, has a vertical face, and the side turned from the water, or the outer face of the wall, a concave face.

2°. That as the height increases, the thickness increases less rapidly, so that in a wall constructed with a vertical face on the water side and a curved face on the other side, and so planned that it shall satisfy the conditions of stability as to its base will present an excess of strength for the surplus of height.

Fig. 7 is the profile of a dam of this description. It will be observed, moreover, that in this form of profile the thickness of the wall at the top is zero This, of course, in practice is never ad-

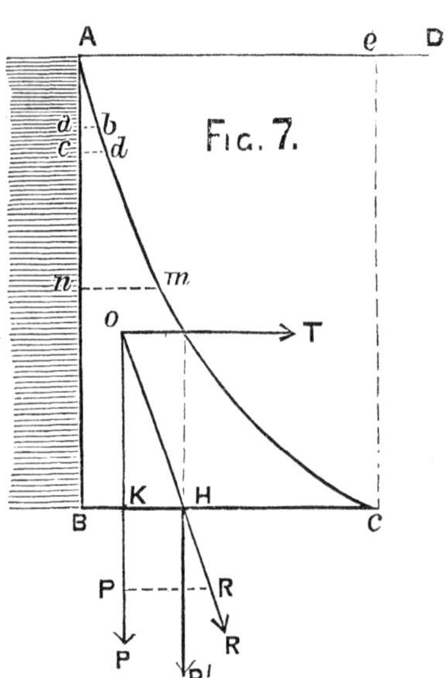

missible, inasmuch as it presupposes the water to be at all times in a perfectly quiescent state, and thus makes no allowance for the very considerable force of the waves raised by the wind. It is, therefore, necessary, whatever the profile, to give the dam quite a thickness at the summit, in general, about fifteen feet, is a good width, as it thus enables us to construct a footpath and roadway on the top of the dam, which is quite a convenience.

Before we consider any other modifications, it may be well to determine as nearly as possible the co-ordinates of the concave curve forming the outer face. For this purpose we will take the vertical face A B as the axis of x, and for the axis of y, a perpendicular to this passing through the point A, and call it AD. Anywhere on the curve we will take a point C, and denote its co-ordinates B C $=y$ and C $e=x$; then the relation existing between x and y will give the equation of the curve. Now, as we have already seen, the wall is subject to the action of two forces, the weight of the dam P, which acts vertically downwards through the centre of gravity and the horizontal thrust T of the water. These two forces produce a resultant R, which cuts the base of the dam in this case at the point H. This resultant, therefore, may be regarded as applied directly to the point H and resolved into two components, H P and H C, respectively, parallel to O P and O T. We have also seen by equation 5 that the horizontal thrust of the water is equal to

$$F' h y = \tfrac{1}{2} h^2 l y \quad . \quad . \quad . \quad 26.$$

Or replacing h by its value, and $l y$ by its value δ, then T, or the horizontal thrust of the water, equals

$$T = \frac{\delta x^2}{2} \quad . \quad . \quad . \quad . \quad 27.$$

And in the same way

$$P = \delta' \int_0^y y\, d x \quad . \quad . \quad . \quad 28.$$

Returning now to equations 24 and 25, we find that the quantity l is equal to y, and that we have therefore to determine the value of u in functions of x and of y. Now u equals H C and H C = K C − K H. The triangles O P R and O K H, moreover, being equiangular triangles are similar, and have their like sides proportional, and

$$KH : PR :: OK : OP$$

or

$$\frac{KH}{PR} = \frac{OK}{OP}$$

or to express the equality in terms of T, x and P,

$$\frac{KH}{T} = \frac{x}{3P} \quad . \quad . \quad . \quad . \quad . \quad . \quad 29.$$

Replacing in the 29th equation the values of T and P, as obtained in the 27th and 28th equations, we have :

$$\frac{KH}{\frac{\delta x^2}{2}} = \frac{x}{3\delta' \int_0^y y\,dx} = KH = \frac{x}{\frac{3\delta' \int_0^y y\,dx}{\frac{\delta x^2}{2}}}$$

$$= \frac{\delta x^3}{6\delta' \int_0^y y\,dx}$$

Or, for brevity, representing $\frac{\delta}{\delta'}$, by D,

$$KH = \frac{D x^3}{6 \int_0^y y\,dx} \quad . \quad . \quad . \quad 30.$$

This gives us the value of K H in the expression

$$u = KC - KH \quad . \quad . \quad . \quad . \quad 31.$$

But K C is evidently equal to $y - BK$, in which B K is the distance from the centre of gravity of the surface A B C to the vertical axis of x or A B. This distance is equal to the sum of the moments of the areas such as $a\,b\,c\,d$, or

$$BK \int_0^y y\,dx = \frac{\int_0^y y^2\,dx}{2} \quad \text{or again,}$$

$$BK = \frac{\int_0^y y^2\,dx}{2\int_0^y y\,dx}$$

Hence $KC = y - BK = y - \dfrac{\int_0^y y^2\,dx}{2\int_0^y y\,dx}$

$$= \frac{2\,y\int_0^y y\,dx - \int_0^y y^2\,dx}{2\int_0^y y\,dx} \quad . \quad . \quad . \quad 32.$$

Substituting in equations 31, the values of K H and K C obtained in equations 30 and 32, we have :

$$u = \frac{2\,y \int_{o}^{y} y\,d\,x - \int_{o}^{y} y^2\,d\,x}{2 \int_{o}^{y} y\,d\,x} - \frac{\mathrm{D}\,x^3}{6 \int_{o}^{y} y\,d\,x}$$

Or, reducing to a common denominator, and subtracting,

$$u = \frac{6\,y \int_{o}^{y} y\,dx - 3 \int_{o}^{y} y^2\,d\,x - \mathrm{D}\,x^3}{6 \int^{y} y\,d\,x} \quad . \quad 33.$$

Thus, then, we have the value of u in functions of x and y, and substituting this value for u in equation 24, and remembering that $l = y$, we have :

$$\frac{4\,y\,\mathrm{P} - 6\,u\,\mathrm{P}}{\delta'\,y^2} = \lambda$$

$$4\,\delta' y \int_{o}^{y} y\,d\,x - 36\,y\,\delta' \int y^2\,d\,x$$

$$\frac{\dfrac{-18\,\delta'\int_0^y y^3\,dx-6\,\mathrm{D}\,x^3\,\delta'\int_0^y y\,dx}{6\int_0^y y\,dx}}{\delta'\,y^2}=\lambda$$

$$\frac{\begin{array}{l}24\,\delta'\,y\int_0^y y^2\,dx-36\,y\,\delta'\int_0^y y^2\,dx\\ \quad+18\,\delta'\int_0^y y^3\,dx+6\,\mathrm{D}\,x^3\,\delta'\int_0^y y\,dx\end{array}}{6\,\delta'\,y^2\int_0^y y\,dx}=\lambda$$

$$-12\,y\,\delta'\int_0^y y^2\,dx+18\,\delta'\int_0^y y^3\,dx$$
$$+6\,\mathrm{D}\,x^3\,\delta'\int\ y\,dx=6\,\delta'\lambda\,y^2\int_0^y y\,dx$$

Dividing both members of the last equation through by $6\,\delta' y\int_0^y y\,dx$, wes hall have, after bringing all terms containing y into the first member,

$$-2\,y\int_0^y y\,dx+3\int_0^y y\,dx+\mathrm{D}x^3-\lambda\,y^2=0 \qquad 34.$$

By making the proper substitutions in equation 25,

$$\left.\frac{2\,\delta'\int_0^y y\,dx}{3\,\delta'\left\{\dfrac{6\,y\int_0^y y\,dx-3\int_0^y y^2 dx-Dx^3}{6\int_0^y y\,dx}\right.}\right\}=\lambda$$

$$2\,\delta'\int_0^y y\,dx=$$

$$\frac{18\,\delta'\lambda\,y\int_0^y y\,dx-9\delta'\lambda\int_0^y y^2\,dx-3\,\delta'\lambda\,Dx^3}{6\int_0^y y\,dx}$$

$$12\,\delta'\left(\int_0^y y^2\,dx\right)^2=$$

$$18\,\delta'\lambda y\int_0^y y\,dx-9\,\delta'\lambda\int_0^y y^2\,dx-3\,\delta'\lambda\,Dx^3$$

Transposing, after dividing each member by $3\,\delta'\lambda$, we have :

$$0=4\left(\int_0^y y^2\,dx\right)^2-6\,\lambda\,y\int_0^y y\,dx+3\,\lambda\int_0^y y^2\,dx+\lambda\,Dx^3\quad .\ .\ .\quad 35.$$

But here a new difficulty presents itself, for no sooner do we attempt to integrate equations 34 and 35, than we see it is quite impossible to perform the integration by any exact method. We may, however, obtain an approximately correct solution by finding the value of y in a series of functions x. Treating equation 34 by this method we obtain, says M. Delocre, for y the value

$$y = a x^{\frac{3}{2}} + b x^{\frac{5}{2}} + c x^{\frac{7}{2}} + d x^{\frac{9}{2}} + e x^{\frac{11}{2}} + f x^{\frac{13}{2}} + \&c. \quad 36.$$

While equation 35 gives :

$$y = a x + b x^2 + c x^3 + d x^4 + e x^5 + f x^6 + \&c. \quad 37.$$

These equations, as it is quite apparent, are of no earthly value for practical purposes, and we shall, therefore, drop all further consideration of them. Indeed, if it were possible to obtain the equations of the curve A m C, by a short and simple process of integration, a moment's reflection will show that such a profile as that illustrated in Fig. 7 would not be suitable for practical use. For this profile has been calculated on the hypothesis that the dam is *always* to support a head of water equal to its height, and in this case the pressure on any horizontal section as $m\,n$ will, it is quite true, not exceed the limit R. But as it happens that the dam is very likely to be at times empty, the profile must be such that, full or empty, the pressure on any section as $m\,n$ shall not be greater than R. We know that this limit will not be exceeded for the face of the wall bounded by A m C, and it thus remains to consider only the vertical face A B. On reference to the calculations we have made relative to the profile of walls having only their own weight to support, it becomes noticeable that the limit will soon be passed if the wall is slightly raised. Supposing this limit to be reached at the point n, we are forced for the sake of stability to depart from the vertical below this point, to give the water face a swelling or bulging surface, and thus adopt a profile similar to that illustrated in Fig. 8. This profile is supposed to fulfill the conditions that, at any section as $d\,e$, taken below $m\,n$, the pressure at the point e, the dam being full, will be less than or equal to the limit R, and the dam being empty, the pressure at *d* re-

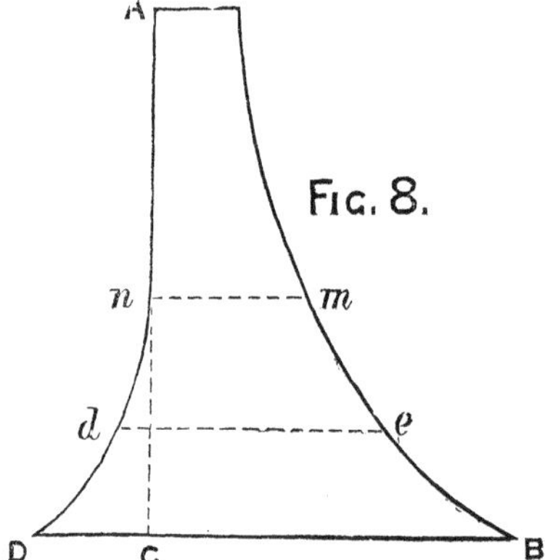

Fig. 8.

sulting from the weight of the structure, will also be less than or equal to the same limit of pressure, R.

This last modification, moreover, is one of no small importance, as it enables us to correct some of the chief errors in which the theoretical consideration has unavoidably led us, and thus to approach nearer to the end in view; the determination of a profile of equal resistance suitable to practical requirements. If the two curves *me*B and *nd*D could be readily obtained by the above formula, the profile of Fig. 8 would answer almost all necessary conditionsas tostabilityand economy; but they cannot. It therefore remains to do the next best thing, and to replace the curved surfaces, by polygonal surfaces of as small sides as possible—in order that they may approach reasonably near to the curves—and then determine the equations of these sides of the polygons; or to adopt a similar method to find the equations of the two curves in question. This we shall now endeavor to do. It is, however, to be remarked that there are two notable instances of the use of the form of profile, shown in Fig. 8; that of the dam at Furens, and that constructed on the Ban, a tributary of the Gier, by M. Mongolfier. Each of these we shall consider later.

As this form of profile, therefore, has been illustrated, and its economy, durability and strength fully tested in the case of the dam at Furens, and in that over the Ban, we shall now undertake its investigation, and determine a series of formulæ for the calculation of the logarithmic curves forming the inner and outer face of the dam, and, finally, the establishment of a *profile type* suitable for dams of various heights. Our investigation, moreover, is to be based on the practical experience of MM. Graeff and Mongolfier, in the construction of the dams of Furens and over the Ban, and the brief but thorough report of Professor Rankine on this form of profile, to many parts of which we are greatly indebted.

In the first place, as to the limit of pressure, two questions naturally present themselves: first, what shall be the greatest limit of pressure we may with safety assume? and secondly, is the same limit to be adopted for the inner as for the outer face of the structure? As regards the first question, it becomes evident at a glance that the limit R′, to which any point in the dam may be subjected without thereby endangering stability, will depend, to no small extent, on the nature of the stone, cement, or mortar used. Yet here, as in other cases where masonry is used, it is possible to assign a general limit, based upon practical experience, which should not in any case be overstepped, and if possible rarely equaled. In the two dams to which we have above alluded, the limit of the pressure was taken at 6 kilogrammes per square centimetre, or 60,000 kilogrammes to the square metre, or taking the kilogramme as equal to 2.20485 pounds, 132.291 lbs. per square metre, which in turn is equal to 1.1954 square yards. In Spain, however, and indeed, we believe in some instances in France, the limit of pressure has been taken so high as 14 kilogrammes per centimetre, and the dam found to stand well, but in the majority of cases at from 6 k. to 8.50k., generally at 6 k., per square centimetre. We may express this pressure in another form much more familiar to English engineers, and take as the limit of pressure for each square foot or square yard, a column of masonry having that area for a base and a height of 160 feet. This is also based on experience, as it is well known that good rubble masonry will, when laid in strong hydraulic cement, bear with safety the pressure arising from the weight of a column 160 feet in height. Taking, again, the density of masonry as double that of water, this pressure would be equaled by a water column 320 feet high, or a pressure per square foot of 20,000 pounds.

The next question as to whether the limit of pressure should be the same, both for the inner and outer face of the dam, seems to be viewed very differently by different engineers, and to admit in practice of a variety of solutions. In the dams constructed by M. Graeff and M. Mongolfier, and in the theoretical profiles offered by M. de Sazilly and M. Delocre, the same limit of pressure was adopted for each face, and the discussion of the formulæ thus much simplified. Yet there seems to be much ground for departing from this observance and for adopting two limits, one for the outer and one for the inner face, provided that the dam has such a logarithmic curve of profile as that we are considering. It is evident that the vertical pressure along these two faces is, at different times, unequal; that when the water is of great depth behind the dam the outer face is more severely strained than the inner, and that when the water is very low, and the dam has little more than its own weight to resist, directly the opposite result takes place and the severest strain is found along the inner face. It is likewise evident that the pressure at any point along these faces must, in all cases, be of necessity in the direction of the tangent to the surface at that place. If the face is vertical, the quantity we derive by the usual equations is the true vertical pressure, or rather the *entire* pressure. But when the surface slopes off from the vertical, as it does in this case, the pressure is in the direction of the tangent, is *inclined* to the vertical, and the quantity which the formula gives us is not the entire pressure, but only its *vertical* component. The *whole* or *real* pressure of course, exceeds this vertical component, by a ratio which grows greater and greater as we pass down the face of the dam to parts where the batter, or slope of the face, departs more and more largely from the vertical. But the outer face has a very much greater batter than the inner, and the water being high, is subjected to a much greater strain, so that, to equalize matters, and not allow the outer face, when the dam is full, to suffer a greater strain than the inner face when the dam is empty, it beccmes most expedient to take a lower limit for the vertical pressure at the outer than we do for the intensity of the vertical pressure at the inner face.

Adopting this view, it remains to fix these two limits of vertical pressure. On the inner face, it is clear, where the slope deviates so very little from the vertical that, for all intents and purposes, it may be safely neglected, we may take that we have already fixed upon, namely, the weight of a column of masonry 160 feet high. For the outer face, we may take a pressure whose vertical component is represented by the weight of a masonry column 120 feet high, a pressure which has been deduced from the practical examples of M. Graeff.

The next matter to be taken into account is that of tension, which must, so far as possible, be avoided in every portion of the dam. And this brings us to the consideration of the "lines of resistance," of which in structures subjected to such varying pressure, there are of necessity two; one for the condition that the dam or reservoir is full of water, and one for the condition that it is empty. As in the case of earth retaining walls and buttresses, these are lines passing through the centre of gravity of each course of masonry, and may, when the faces of the dam are rectilinear, be found by any of the formulæ used for such purposes. They bear, therefore, intimate relations to the stability of the dam, the latter decreasing as they depart from the centre of thickness and near the faces. They also bear relation to the tension, and in order that the latter may not become appreciable in any part of the structure, they must not deviate at any point from the line passing through the centres of thickness, either outward when the dam is full, or inward when empty, by a distance greater than one-sixth of the thickness at that point.

With these conditions in view, we now pass to the consideration of the profile.

PROFILE TYPE FOR DAMS HAVING CURVES FOR BOUNDING FACES.

Let Fig. 9 represent the profile of a dam bound by logarithmic curves, the various equations relative to which we wish to find. Let the vertical line A S represent the asymptote of the curves, and taking the origin of co-ordinates at the top of the dam, represent by x all horizontal, and by y all vertical measurements, by b the breadth or thickness of the dam across the top,

A
18
10
30
50
70
90
100
120
140
60
133
RESERVOIR FULL.
RESERVOIR EMPTY.
S

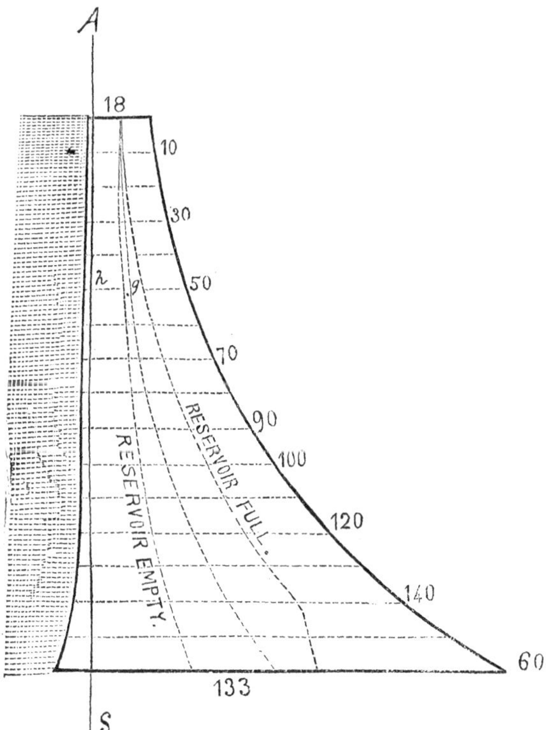

FIG. 9.

and by b' the breadth at any other place lower down. Also let s represent the sub-tangent common to the two curves, and represented in the figure by that part of the asymptote contained between F and G. As to the lines of resistance let their deviation from the middle of the thickness when the dam is full and empty be expressed by the letters r and r' respectively, and by R and R′ denote the limits of pressure ; the first for the outer, the second for the inner face.

Now, adopting Professor Rankine's method of procedure, it becomes evident that if the thickness across the top be expressed by b, then the thickness at any other portion of the dam lower down, and at a distance y below the top, will be expressed by the equation

$$b'=b.e^{\frac{y}{s}} \quad . \quad . \quad . \quad . \quad . \quad 38.$$

in which e is the modulus of the common system of logarithms, or 0.434294. To apply this equation therefore to practice, it is necessary to know the value of the sub-tangent, the thickness across the top and the vertical distances of different points on the face of the dam below the axis of X. These latter points are, of course, assumed at random, and have in the present case been taken five feet apart. As to the thickness at the top it has been taken at eighteen feet. In the dams already alluded to (those of MM. Sazilly and Mongolfier) with the height of 50 and 42 metres respectively, and a limit of pressure of 60,000 kilogrammes per square metre, the thickness across the top is, in the former, five, and in latter, five and seven-tenths metres, which, expressed in feet, gives for the one 16.4 and for the other 18.6 feet. But in this instance we have slightly enlarged on the thicknesses used by those engineers, in order to produce a profile suited for a dam required to resist not only the thrust of water, but also that of ice when carried down by spring freshets. The determination of the sub-tangent s is not so obvious, but may be found by giving to the exponent $\frac{y}{s}$ of e an approximate value of , which substituted in the formula of Prof. Rankine, gives a corrected value of $\frac{11}{8}$, and a sub-tangent equal to 80 feet.

If, then, adopting this breadth of 18 feet on top, we desire to find that at a point thirty feet below, we may write equation :

$$b' = \log. b + 0.434294 \times \frac{30}{80} \quad . \quad 39.$$

$$= 1.255273 + 0.162858 = 26.19 \text{ feet},$$

which is to be measured off in such wise that thirteen-fourteenths of it shall lie on the down stream or outer side of the asymptote, and the remaining one-fourteenth on the up stream or inner side. Taking other values for y and proceeding in precisely the same way, we thus obtain any desired number of points through which must pass the logarithmic curves that form the faces of the dam. This done and the curve drawn, the next step is to determine the lines of resistance when the dam is full and when it is empty. To begin with the latter case, the dam being empty, the deviation of the line of resistance from the middle of the thickness will evidently be *inward* or towards the up stream side of the dam. This deviation we have expressed by the letter r', and if we wish to find its value for a horizontal section of the dam taken 50 feet below the top, we proceed as follows. Let z denote the distance $h\,g$ or the deviation of the centre line of the thickness outward from the axis A S, and by z' the deviation of the same line from the same axis at the top of the dam. Referring to Fig. 9, the distance we wish to find is evidently equal to $g\,h$ minus the deviation of the centre of thickness of the top of the dam from A S, divided by 2, or

$$r' = \frac{z - z'}{2} \quad . \quad . \quad . \quad . \quad 40.$$

Because the dam having only its own weight to carry, the line of resistance must cut the line $g\,h$ in a point vertically below the centre of gravity of that part of the structure above $g\,h$.

The thickness of the dam where y is fifty feet is found from equation 39 to be 33.63 feet; the centre of thickness 16.81, and the value of z or the deviation of this centre from the axis A S is 14.41 feet. That of z' or the deviation at the summit of the dam is 7.72 feet, from which it follows that (eq. 40) $r' =$ 3.35 feet. It is in this way that the values of r', given below in Table A, have been calculated.

It is next necessary to determine an equation from which to find the values of r, or the amount by which the line of resistance deviates outward from the centre of thickness when the dam is full. It is evident this deviation will depend upon three things, the moment of the horizontal thrust of the water, above the section at which we wish to find r, the weight of the dam above this same section, and the amount by which the line of resistance is moved *inward* when the dam has only its own weight to carry, so that if we divide the moment of the thrust by the weight, and *subtract* the quantity r', we shall at once have the value of r. The thrust of the water above any horizontal section of the dam is, as we have already seen by equation 2, $\frac{y^2}{2} \times 62.5$ lbs., and the moment is, therefore, $\frac{y^2}{2} \times 62.5 \times \frac{y}{3} = \frac{y^3}{6}$ 62.5 lbs., or, what is the same thing, if we express by w the ratio in which the masonry is heavier than the water, and take, as is usual, this ratio as 2, we shall have for the moment (expressed by m) of the horizontal thrust of the water,

$$m = \frac{y^3}{6\,w} = \frac{y^3}{12} \quad . \quad . \quad . \quad . \quad 41.$$

The weight of any lineal unit of the dam above the section may be found most simply by the calculus. Thus giving to y and b the same signification as before, and taking the weight of a cubic unit of masonry as the unit of weight, the weight of each unit of length of the wall above the section is expressed by

$$W = \int_0^y b'\,dy \quad . \quad . \quad . \quad 42.$$

Integrating this between the limits y and o, and remembering that $b' = b\,e^{\frac{y}{s}}$ we have :

$$W = s\,b\left(e^{\frac{y}{s}} - 1\right) = s\left(b \, . \, e^{\frac{y}{s}} - b\right)$$

$$= s\,(b' - b) \quad . \quad . \quad . \quad 43.$$

For r, therefore, we have :

$$r + r' = \frac{m}{W}$$

$$r = \frac{m}{W} - r' = \frac{\frac{y^3}{12}}{s\,(b' - b)} - r'$$

or

$$r = \frac{y^3}{12\,s\,(b' - b)} - \frac{z - z'}{2} \quad . \quad 44.$$

This equation gives for the value of r at the distance fifty feet below the top, the quantity 5.18 feet, which, as it falls below one-sixth of the thickness at this point, we are justified in considering the deviation as not too great to be perfectly consistent with stability.

But, to make assurance doubly sure, we may apply a final test as to stability, by calculating the amount of vertical pressure at various points along both the inner and outer faces, and comparing the results with the limit of pressure, which, it will be remembered, has been fixed for the inner face at weight of a column of masonry 160 feet in height, and for the outer face at that of a column 120 feet high. This matter we have already considered at length, and have deduced two equations, 13 and 14, which as they are perfectly suited to the present case, we shall not delay to deduce others, but alter them to suit the notation of Fig. 9. Thus altered they are, calling p and p' the pressures at the outer and inner face respectively, and P and P′ the limit at these same faces—

$$\left.\begin{array}{l} p=2\left(2-\frac{3u}{b'}\right)\frac{W}{b'} = \text{ or } < P \\ \text{and} \\ p=\frac{2W}{3u} = \text{ or } < P \end{array}\right\} \quad 45.$$

While for p' we have two others precisely similar, with the exception that P in equation 45 is changed to P′. It may, perhaps, be well to again remark that the first or second value of p in equtaion 45 is to be used according as the value of u is greater or less than one-third of the thickness, and that in all such profiles as that of Fig. 9, the quantity u denotes the distance from *the outer face* to the *line of resistance when the dam supports a charge of water*, and from the *inner face* to the *line of resistance when the dam or reservoir is empty.* To illustrate by one example, let it be required to find the vertical pressure at the point C, on the outer face of the dam (Fig. 9), situated fifty feet below the top. By referring to Table A, we see that b' is equal to 33.63 feet, that the outward deviation of the line of resistance is 4.98 feet, and that u must therefore be 11.83 feet. The quantity $W = s(b' - b)$ is 1250.4. Since u is here greater than $\frac{b'}{3} = 11.21$, we use the first of equation 45, and, making the substitution of values, we have :

$$p=2\left(2-\frac{35.49}{33.63}\right)\frac{1250.4}{33.63}=70.555$$

Thus showing that the pressure is but a little more than half the limiting pressure. Precisely the same operation repeated, with u equal to 13.46 feet, will give the amount of vertical pressure at the inner face at a point fifty feet below the top, the dam supporting only its own weight. This pressure is thus found to be equal to a column of masonry 59.4 feet in height.

The area of the entire profile or of any portion of it, included between two horizontal sections, may be found by taking the difference between the thickness of the dam at these two sections, and multiplying the difference by the subtangent. For it is evident from the figure that, if b equals the thickness of a point y feet from the top, then this thickness multiplied by the differential of the height and integrated between the limits y and zero, is the area, and this expression $\int_0^y b'\,dy$ when integrated, remembering that b' is equal to $b e^{\frac{y}{s}}$ gives $s b e^{\frac{y}{s}} - s b$, or replacing $b e^{\frac{y}{s}}$ by b', the expression for the area becomes $s(b'-b)$. In the notation we have used b means the thickness of the dam across the top, but in calculating the area of any portion of the profile not bounded by the top thickness, the quantity b is to be understood to mean the smaller of the two thicknesses which bound the area. That is to say, if we wish to find the area of that portion of the profile included between horizontal sections taken at thirty and eighty feet below the top, b represents the thickness at the former section, and we have 80 (48.93 −26.19) = 1819.2 square feet. Having the area, the solid contents and weight for any length of the dam are of course readily found. The areas for sixteen different sections of the profile, each having the top of the dam for one side, have been calculated in this way, and will be found entered in the last column of Table A. The first column of this table gives the distances in feet of the sections estimated from the top downwards, the second the thickness of the dam at these sections, the third the deviation of the line of resistance outward when the reservoir is full, the fourth the deviation inward when empty, and the last the areas.

TABLE A.

		r'	r	Area sq. feet.
0	18.00	0	0	°
10	20.40	.51	.18	192.00
20	23.10	1.09	.54	408.00
30	26.19	1.75	1.68	655.20
40	29.68	2.52	3.18	934.40
50	33.68	3.35	5.18	1254.40
60	38.10	4.53	6.66	1608.00
70	43.17	5.39	8.79	2013.60
80	48.93	6.62	10.52	2474.40
90	54.18	7.75	12.95	2894.40
100	62.97	9.63	13.53	3597.60
110	71.18	11.39	15.02	4254.40
120	81.79	13.62	14.59	5103.20
130	91.39	15.72	15.46	5871.20
140	103.60	18.34	15.05	6848.80
150	115.00	20.78	15.46	7440.00
160	133.00	24.64	12.46	9200.00

It is perhaps unnecessary to call attention to the fact, that this form of profile has been calculated with a view to its serving as a *profile type* for dams of any height, great or small, whose faces are logarithmic curves. For a dam, then, of which the height is thirty feet, that portion of Fig 9, above the line marked 30, is the proper profile : for one eighty feet in height, that portion above the line marked 80, and so for each succeeding section. It presents again many strong points not found in dams of the usual rectilinear profile, which are especially deserving of consideration when damming a river or valley of great breath and depth. Of these not the least is its economy of material, which, as we shall hereafter see, is very great as compared with that of stepped or sloping profiles ; while the curves of the two faces are so gradual that no great mechanical difficulty can arise in cutting the facings. Another matter, which, in the dams of Furens and the Ban was not taken into account, that of tension, has here been considered and the profile so determined that when the reservoir is full the tension on the outer face shall not at any point be greater than it is on the inner face when empty.

The profile of the Furens dam is given in Fig. 10, and that constructed on the

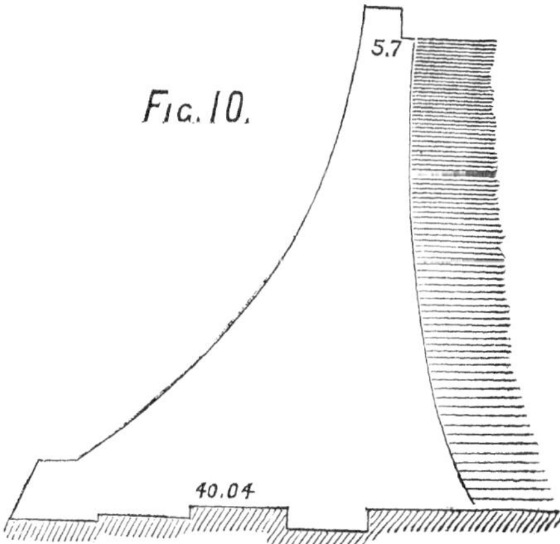

FIG. 10.

Ban, a tributary of the Gier, in Fig. 11. The former has a height of fifty metres with a breadth on top of 5.70 metres,

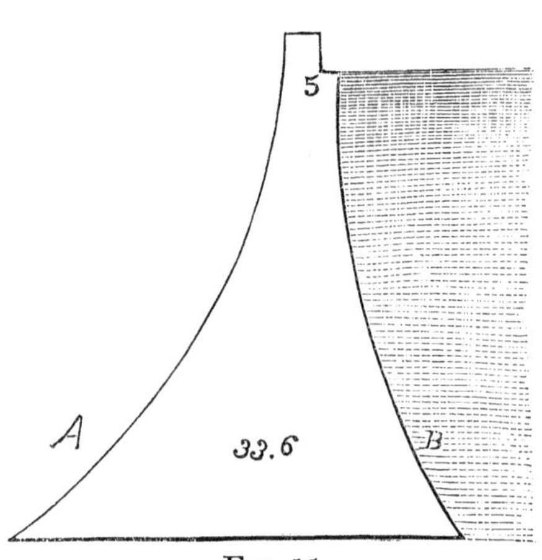

FIG. 11.

and a limit of pressure of six kilogrammes per square centimetre. The latter has a height of forty-two metres, a thickness on top of five metres, with the same limit of pressure as the Furens dam. By a comparison however, of the profile of the former with that part of the profile of the Furens which lies above the limit A B we see that the thickness has been very considerably reduced, while if we extend the profile to fifty metres and then compare it with the Furens, we find that the pressure nowhere exceeds 8 kilogrammes to the square centimetre.

To return now to the modifications of which this type of profile is susceptible.

MODIFICATIONS OF THE LOGARITHMIC PROFILE.

On a moments inspection of Fig. 8, it is readily seen that, as the inner curve does not anywhere depart very far from the asymptot A S, the first and simplest modification of this curve is to replace it by a right line and thus make the inner face vertical from top to bottom. But the outer curve if treated in like manner, and replaced by a right line, would give us a form of profile which, though it possessed no more thickness at the bottom than was absolutely necessary to withstand the vertical pressure, would at every other point, possess a thickness greatly in excess of the requisite amount, and thus occasion a prodigious waste of masonry. We must therefore, break this continuous slope and substitute for one long line two or more shorter ones each of which makes a different angle with the vertical. Limiting our attention for the present to the first case, and replacing the two logarithmic curves in Fig. 9 by lines,—the inner curve by one vertical, and the outer by two inclined—we have produced for us a profile of the form illustrated in Figs. 12 and 13. The question that first presents itself in the discussion of such a profile, is evidently how far down the outer face the point C is to be taken. It comprises indeed, the *entire* discussion. Of course, it is a great advantage, so far as the saving of material is concerned, to throw this point as low as possible, but this is limited by the condition, so necessary to secure stability, that when the reservoir is full the vertical pressure at C shall not be *greater* than the limiting quantity R. Having determined the thickness across the top, which preserving our previous notation, we will call b, the quantities to be determined are first, the vertical distance of the point C below the top, and

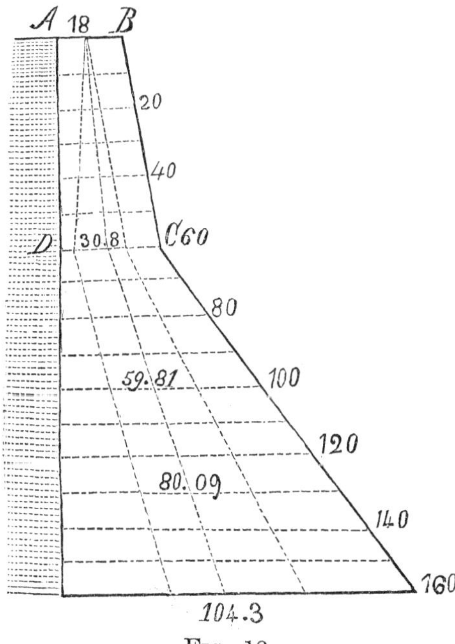

Fig. 12.

second the thickness of the dam at this point, or what is perhaps more easily obtained the *excess* of the thickness at C over the thickness at the top, A B. The distance, A D (Fig. 12) we will call y;

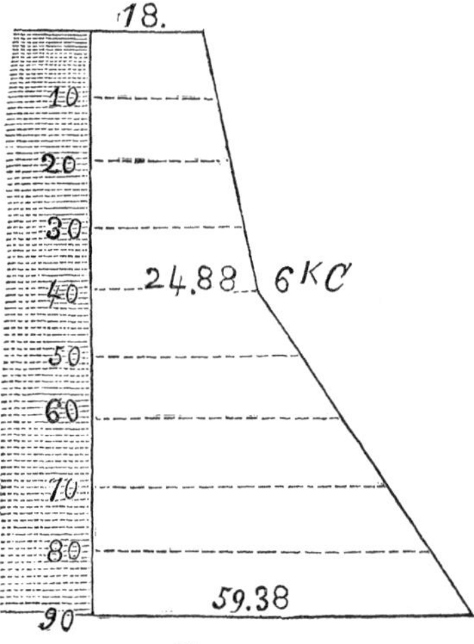

Fig. 13.

the total thickness D C we will represent by b', and express the excess of thickness by v. By W, denote the weight of the part A B C D (Fig. 14), and by F, the horizontal thrust of the water above D. These two forces act through the centre of gravity O, the former vertically downward and represented in Fig. 14 by

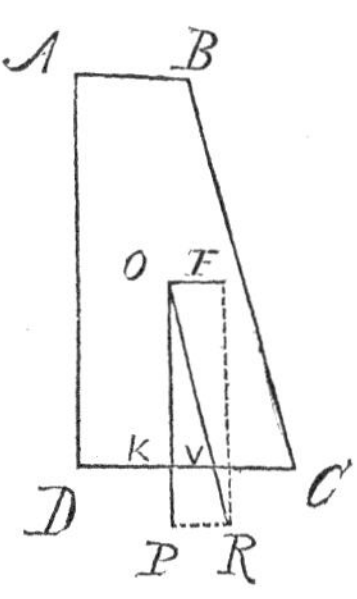

FIG. 14.

the line O P ; the latter horizontally and represented in direction and intensity by O F. These two produce a resultant which cuts the base at V, and this point may therefore be regarded as the point of application. From this relation, as we have seen, result two equations $2\left(2-\frac{3u}{l}\right)\frac{P}{l}=$ or $<R'$, and $\frac{2P}{3u}=$ or $<R'$, which are to be used according as u is $>\frac{l}{3}$ or $<\frac{l}{3}$.

In these equations $P = W$, is, according to the notation of Fig. 14, expressed by $\left(\frac{b+b+v}{2}\right) y \delta'$, in which δ' is the density of the masonry; $l=b'=b+v$ and $u=$ C V. But $u=$ C V $=$ K C $-$ K V. Now K C may be found by the equation expressing the relation that the moment of the weight of A B C D, with respect to C, is equal to the sum of the moments of the two parts A B V D and B V C into which the area of A B C D may be divided. The moment of the weight of A B C D, with respect to C, is evidently the weight $\left(\frac{2b+v}{2}\right) y \delta'$ multiplied by K C; that of A B V D by $\frac{(b+2v)\, b\, y\, \delta'}{2}$ and that of B V C by $\frac{y v^2 \delta'}{3}$. Hence, the relation when expressed, becomes:

$$\text{K C} \times \left(\frac{2b+v}{2}\right) y \delta' = \frac{(b+2v)\, b\, y\, \delta'}{2} + \frac{y\, v^2\, \delta'}{3} \quad . \quad . \quad . \quad . \quad 46.$$

$$\text{K C} = \frac{\frac{(b+2v)\, b\, y\, \delta'}{2} + \frac{y\, v^2\, \delta'}{3}}{\left(\frac{2b+v}{2}\right) y\, \delta'}$$

$$= \frac{3\,b^2+6\,b\,v+2\,v^2}{6\,b+3\,v} \quad . \quad 47.$$

To find K V, we have from the two similar triangles O K V and O P R the proportion

$$K\,V : K\,O :: P\,R : P\,O$$

whence

$$K\,V = \frac{K\,O \times P\,R}{3\,P} \text{ or } \quad . \quad . \quad 48.$$

since P R is equal to the horizontal thrust, which, as we see in the early part of our investigations, is equal to $\frac{y^2\,\delta}{2}$; and since P O is equal to the vertical pressure and this is equal to $\left(\frac{2\,b+v}{2}\right)\,y\delta'$ we have finally for the value of K V :

$$K\,V = \frac{y^2\,\delta}{3\,(2\,b+v)\,\delta'} \text{ or } K\,V = \frac{y^2\,\theta}{3\,(2\,b=v)} \qquad 49.$$

which latter equation is found by substituting for $\frac{\delta}{\delta'}$ the letter θ. These values given in equations 49 and 47 when replaced in the expression

$$u = KC - KV$$

give $u = \frac{3b + 6bv + 2v^2}{6b + 3v} - \frac{y^2 \theta}{3(2b + v)}$

$$= \frac{2v(v + 3b) + 3v^2 - \theta y^2}{6b + 3v} \quad . \quad 50.$$

With this value of u we return to equations 24 and 25, and, substituting it, we obtain :

$$2\left\{\frac{2 - \frac{6v^2 + 18bv + 9v^2 - 3\theta y^2}{6b + 3v}}{b + v}\right\} \times \frac{\left(\frac{2b + v}{2}\right) y \delta'}{\delta' \lambda} = \lambda$$

and $\dfrac{2\left(\frac{2b + v}{2}\right) y \delta'}{3\delta'\left(\frac{2v^2 + 3b + 3v^2 - \theta y^2}{6b + 3v}\right)} = \lambda$

These, when reduced and made equal to zero, give us two equations containing two unknown qualities :

$$\theta y^3 - \lambda v^2 - 2b\lambda v + b^2 y - \lambda b^2 = 0 \quad . \quad 51.$$

$$v^2 y - 2\lambda v^2 - 4bvy + \theta\lambda y^2 - 6b\lambda v + 4b^2 y - 3b^2\lambda = 0 \quad . \quad . \quad 52.$$

The first of which is to be used when $u > \frac{b}{3}$, and the second when $u < \frac{b}{3}$. Each of these equations express the relation that when the reservoir is full the vertical pressure at the point C (Fig. 14) shall be equal to the limit R. But we must also take into consideration the inner face, and find an equation expressing the relation that the reservoir being empty, the pressure at D, shall not exceed the limit R. In this case, the face being vertical, the pressure of the water does not exist, and the force P, or the weight of this portion of the dam, acts downwards through the centre of gravity, and

$$u = \mathrm{D\,K} = \mathrm{D\,C} - \mathrm{C\,K}$$

$$u = b + v - \frac{2\,v\,(v + 3\,b) + 3\,b^2}{6\,b' + 3\,v}$$

$$= \frac{6\,b^2 + 6\,bv + 3\,bv + 3\,v^2 - 2\,v^2 - 6\,bv + 3\,b^2}{6\,b + 3\,v}$$

$$= \frac{v\,(v + 3\,b) + 3\,b^2}{3\,(2\,b + v)} \quad \ldots\ldots\ldots \quad 53.$$

With this value of u, we again return

to equations 24 and 25, substitute in each, and reducing, have :

$$v^2 y - v^2 \lambda + 3 b y v - 2 b \lambda v + b^2 y - \lambda b^2 = 0 \quad 54.$$

$$v^2 y - v^2 \lambda + 4 b y v - 3 b \lambda v + 4 b^2 y - 3 b^2 \lambda = 0 \quad 55.$$

By combining 51 and 54, or 52 and 55, we may readily obtain the value of y and v, which are the two quantities we wish to find. It is moreover to be remarked that λ in the above equations is found by dividing the limit of vertical pressure at C and D by the ratio in which the masonry is heavier than water. Thus in calculating the profile of Fig. 12, we have first reduced the limit of vertical pressure per unit of surface from pounds to kilogrammes, and taking the density of water, as given in the French tables, as 1000 kilogrammes and the density of masonry as double that of water or 2000 kilogrammes, we have $\lambda = \frac{R}{2{,}000} = \frac{60{,}000\,k}{2{,}000\,k}$ or $\lambda = 30$. We thus obtain for λ, a very simple number, whereas had we retained the pressure as expressed in pounds, we would have had a much larger one to handle. In Fig. 13 however, in order to produee a profile of what may be considered as a type of the greatest boldness consistent with safety, we have taken the limit of vertical pressure at 14 kilogrammes per square centimetre, which as we have already stated has been used in several instances in France and Spain. This increases the value of λ to 70. The thickness across the top is in each case the same as in that of the profile illustrated in Fig. 9; namely, eighteen feet, but the height of that in Fig. 12 has been reduced to ninety feet. The height AD of the upper part A B C D and the value of v corresponding to it have been found by combining equations 51 and 54. The lower part, by the same equation, by substituting for y the difference between the height A D of the upper part and the entire height of the dam.

The deviation of the line of resistance

when the reservoir is full may also be found as follows. Let A B C D in Fig. 15, represent either the upper or lower

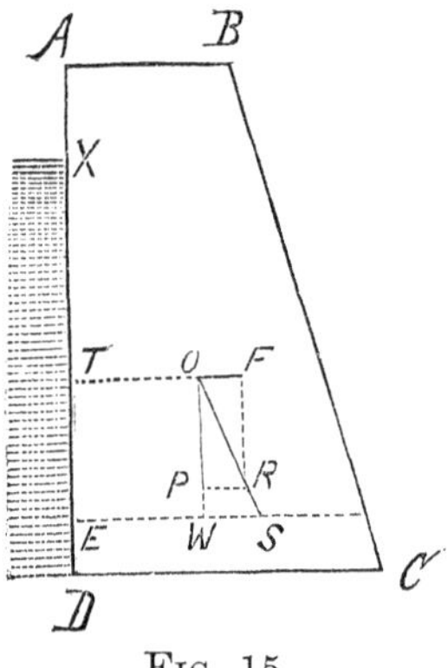

FIG. 15.

part of the dam whose profile is given in Fig. 13, and let it be desired to find the amount of deviation at any section as E F. By O represent the centre of gravity of A B F E, then will O R represent the resultant of the two forces acting on this portion of the dam, and the distance we wish to find will be E S. We will suppose also, in order to cover all cases that the water stands at X. Also let $AB = b$; $AX = l$. $AE = y$; $ES = x$; $EW = h$; and the inclination of the sloping side B C, to the vertical be denoted by $\propto$; by δ the density of the masonry and by δ' that of the water. Then by the similar triangles O P R and O W S, we have :

$$\frac{WS}{OW} = \frac{PR}{OP}$$

Now $WS = ES - EW = x - h$ and $OW = TE = \frac{1}{3} XE$ because the centre of pressure (T) of a rectangular plane surface sustaining the pressure of water, is at a point two-thirds the depth of its immersion. Hence $TE = \frac{1}{3}(y - l)$. P R or the horizontal thrust of the water on X E is, as we know, expressed by $\frac{\delta'(y-l)^2}{2}$; and the pressure O P by

$$\frac{(b+b')y\delta}{2} \text{ or } \frac{(2b + y \tan. \propto)y\delta}{2}$$

$$\therefore \frac{x-h}{\frac{1}{3}(y-l)} = \frac{\frac{1}{2}\delta'(y-l)^2}{\frac{1}{2}y\delta(2by+2y^2 \tan. \propto)\delta} \qquad 56.$$

Then replacing $\frac{\delta'}{\delta}$ by θ

$$x - h = \left(\frac{(y - l)^2}{2\, by + y^2 \tan. \alpha} \right) \frac{1}{3\, \theta} \quad . \quad 57.$$

$$\text{But } h = \frac{\frac{1}{3} y^3 \tan.^2 \alpha + b y^2 \tan. \alpha + b^2 y}{2\, b y + y^2 \tan. \alpha}$$

which, added to equation 56, gives :

$$x = \frac{\frac{1}{3\,\theta}(y-l)^2 + \frac{1}{3} y^3 \tan.^2 \alpha + b y^2 \tan. \alpha + b^2 y}{2\, b\, y + y^2 \tan. \alpha} \qquad 58.$$

This value of x is, of course, to be measured off *from* the vertical side. When the water stands at the top of the dam, the value of l, is zero, but when the reservoir is empty, then l, is equal to the entire height of the dam. The simplest way, however, to find the deviation, is by means of Equation 50, observing that the value of u, when found is to be laid off from the outer or sloping face of the dam ; and corresponds to the distance F S in Fig. 15.

The second modification, then, of the theoretical profile of equal resistance, consists in replacing the outer curved face by a broken one composed of two planes inclined at different angles to the horizon. The principles, however, which justify us in the use of such a modification, may be carried still further, and the inner and vertical face replaced by one almost a fac simile of the outer broken one. Indeed the only essential difference between them lies in the degree of slope which we give to their two plane surfaces. On the one side both are sloping; on the other that portion of the face from the summit of the dam to a point below, (where the pressure on each unit of surface equals the assumed limit of pressure,) the wall is vertical, and from here to the base slope outward. This latter point moreover, must be directly opposite that point on the outer face at which the two sloping lines of the profile intersect. Of a profile thus constructed, some idea may be had from the sixteenth figure. It does not present any merit either as to beauty, strength, stability or economy of material not possessed by that illustrated in Figs. 12 and 13. As to economy indeed, the amount of material consumed is if anything greater in former than in the two latter forms of dams, and it may be justly doubted whether the additional stability thus obtained, is a fair recompense for the additional outlay for material and for cutting facing stones for a third sloping face.

As to the mathematical calculations of such a profile they are rather lengthy than difficult. For the upper portion A B C D, Fig. 16, we have already discussed the principles at length, and obtained in equations 51 to 55 the necessary formulæ. The value of A B or b is of course known, as also that of A D or a' which is assumed, and is not to be greater than λ or the greatest height we can with safety give to a wall with vertical faces. That of the lower portion C D E F, may also be conducted on the principles previously laid down, and as it necessitates several eliminations of somewhat startling length we shall consider it

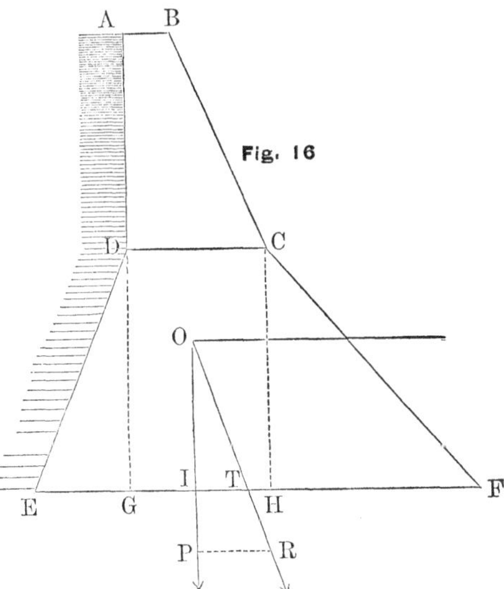

merely in outline. Knowing the total height of the dam, and the distance A D, we of course know D G, or the height of that portion of the dam C D E F, whose breadth of base E F, we wish to find. We also know from equations 51 and 54, the breadth D C, and projecting this on the base we at once obtain that portion of it between G and H. What there remains to be found is G E, and H F. The former of these unknown quantities we will denote by y, and the latter by z; the breadth E F, of the base by b,″ the part G H, which is also equal to C D, by b'; the height D G, of the lower section of the dam by a, and that of the upper section, or A D, by a'. Returning now to the equations 15 and 16, which are the general equations of stability for a dam supporting the pressure of a head of water, we find that the three unknown quantities for which we wish to find values in term of the known quantities we possess are u, l, and p. The value of l, or the thickness E F, of the base is, when expressed in terms of the above notation.

$$l=y+b+z$$

While P is of course the area of the irregular polygon A B C F E D multiplied by the weight per unit of volume, *plus* the *vertical* component of the weight of the water resting on the sloping face

D E. The area of A B C D is $\left(\frac{b+b'}{2}\right)$ $\delta' a'$. That of C D E F is $\frac{y a \delta'}{2}+\frac{z a \delta'}{2}$ $+b' \delta' a$. The vertical thrust of the water is by equation (1), $\left(\frac{2 a'+a}{2}\right) y \delta$. The value of P, therefore, is $b' \delta' a +$ $\frac{y a \delta'}{2}+\frac{z a \delta'}{2}+\left(\frac{b+b'}{2}\right)\delta' a'+\left(\frac{2 a'+a}{2}\right)$ $y \delta$, which reduces to the form

$$P=\frac{2 b' \delta' a+\delta' a (y+z)+2 (b+b') \delta' a'+(2 a'+a) y \delta}{2} \quad 59.$$

Again, to find the value of u, the first step is to construct the diagram of forces, as illustrated in the figure, O P representing in direction and intensity the vertical component P, or the weight of the dam and the water, and O F the horizontal component or the outward thrust of the water behind the dam. Then will F T represent u which is clearly equal to

$$u=z+\text{H I}-\text{I T} \quad . \quad . \quad 60.$$

But by the two similar triangles we have, as before, $IT = OI \times \frac{OF}{OP}$ or since $OI = \frac{a+a'}{3}$ and O F (equation 2) equals $\left(\frac{a+a'}{2}\right)^2 \delta$

$$IT = \frac{(a+a')^3 \delta}{3 \left[2 b'\delta' a + \delta' a (y+z) + 2 (b+b')\delta' a' + (2 a' + a) y \delta\right]}$$

H I is to be obtained in precisely the same manner as K C was obtained from Fig. 14, by expressing the relation that the moment of weight P (which includes, it is to be remembered, that of the dam and that of the water pressing on the inclined face D E), with respect to the point F is equal to the sum of the moments of the components of this force. Obtaining these moments in the same manner as we obtained those for the equations deduced from Fig. 14, and putting them equal to the expression $P \times IF$, or $P \times (IH + z)$, we have after reduction, the equation

$$H = \frac{12 \propto \beta + b'^2 a + 2\, a\, y^2 + 6 b' a\, y + (b\, a' + 3\, a)(y + 2\, b')\, y\, \theta - 2\, z^2\, a}{(12 \propto + b' a) + 6\, a\, z + 6 a y + 12\, a' y \theta + 6 a y \theta}$$

In which $\propto$ is a short expression for the area of A B C D, and β the distance from C to the point where the perpendicular of the centre of gravity of A B C D cuts C D, and this replaced in equation 60, gives for the value of u

$$u = \frac{12 z(\propto + b' a) + 6\, z\, a (y + z) + 6\, z y \theta (2 a' + a) + 12 \propto \beta + 6\, b^2\, a + 2\, a\, y\, (y + 3\, b') + 3\, (2\, a' + a)\, (y + 2\, b')\, y\, \theta - 2\, a\, z^2 - 2\, \theta\, (a' + a)^3}{12(\propto + b' a) + 6\, a\, z + 6 a y + 12 a' y \theta + 6 a y \theta}$$

Eq. 61.

The quantities P, u and l, being thus obtained in terms of b', y, z, a and a', a substitution in equations 15 and 16, will furnish us with two equations of great length, from which, by the process of elimination, the values of x and y are readily found.

To take but one example of this form

of profile, let it be required to calculate the dimensions of such a profile for a masonry dam one hundred and seventy feet in height and eighteen feet broad on top, the limit of pressure being taken at 132,000 pounds. For this purpose we have to determine beforehand the height a' of the part A B C D. This, in the present case, is taken at 80 feet, and may in all cases be assumed arbitrarily. Now, since the dam has one vertical face, we have to determine but one quantity v, or the difference between the thickness of the dam at A B and that at C D, and this value of v is readily obtained from equation 51, which, modified to suit the present notation, becomes

$$\theta a'^3 - \lambda v^2 - 2 b \lambda v + b^2 a' - \lambda b^2 = 0 \quad . \quad 62.$$

Solving this with reference to v, we have

$$v^2 + 2 b v = \frac{b^2 a' + \theta a'^3}{\lambda} - b^2$$

$$v = \sqrt{\frac{b^2 a' + \theta a'^3}{\lambda}} - b \quad . \quad . \quad . \quad 63$$

And replacing the quantities by their

values, remembering that λ equals 98.4 ft., and θ (or the ratio in which the masonry is heavier than water) equals $\frac{1}{2}$, the result finally obtained is,

$$v=53.52-18 \text{ or}$$
$$b'=b+v=53.52 \text{ feet.}$$

With this value of b' we return to the equations expressing the values of x and y as deduced from equations 15 and 16, after the substitution of the value of u given in equation 61, and find that the value of $b''=x+b'+y$ is 178.42 feet.

Once more, we may carry this principle one step further and produce a profile which is little more than a modification of that given in Fig. 16. If, for instance, while preserving the same height of structure, we divide each of the three sloping faces into two parts, and give to each part thus produced a face inclined to the horizon, we shall then have a profile of such shape as that illustrated in the seventeenth figure.

A glance at this is sufficient to show that it is in reality but a compound of the

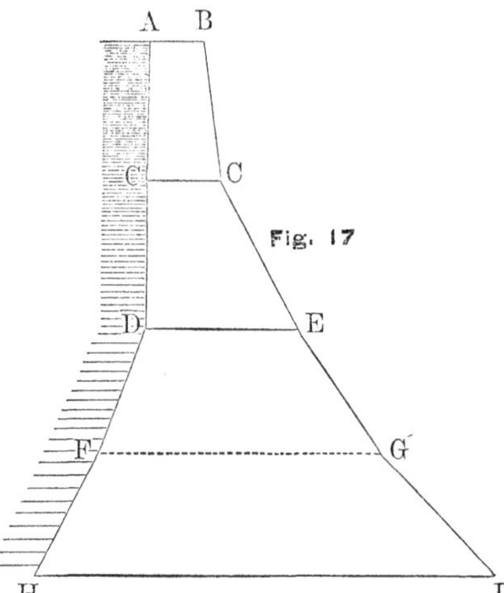

two preceding profiles, and that therefore the principles to be observed in the calculation of its parts are those already discussed. The entire profile may thus be considered as divided into three pieces ;—that from A to D, in which the inner face is vertical throughout, and the outer made up of two inclined faces, constituting a profile exactly similar in design to that of Fig. 12: that from D to F, and that from F to H, in each of which both the outer and inner faces are sloping. The first part is, therefore, to be calculated in the same manner as we would calculate the thickness of a dam having the profile of Fig. 12, and each of the two remaining portions by the equations deduced from Fig. 16. To illustrate this by a case in point, let it be required to find the thickness at various points of a masonry dam, having such a profile as that we are discussing, its thickness across the top being 18 feet, and the total height 170 feet. The first thing that claims attention is the determination of the vertical distances between the points B and C ; C and E ; E and G ; and finally G and I. These may, of course, be chosen at pleasure, just as we may select the number of parts that each face is to be composed of, and as in the present case the dam is 170 feet high, and the outer face divided into four parts, we will for convenience divide the dam first into two equal parts, then divide the lower of these again into two equal parts, and the upper also into two, but two unequal parts. The vertical distances between the sections will then be, beginning at the bottom and going up G I = 42.5 feet; E G = 42.5 feet; C E = 45; and B C = 40 feet. Had the dam, however, been one hundred and fifty, or one hundred and eighty feet high, or indeed any other number, then the best arrangement would again have been, to make the second vertical distance—that from C to D—longer than the remaining three, so that, if the dam was one hundred and fifty feet high, the best arrangement would be B C = 30; C E = 60; and E G and G I each thirty feet; if the height had been one hundred and eighty feet, then B C = 40; C E = 50; and the others each forty-five feet. Although this arrangement may seem to be somewhat arbitrary, it is in reality based upon fixed principles, which clearly show that where such a number of divisions and such a profile as that used in the present instance are employed, the second part should be decidedly longer than either of the other three. Those portions, moreover, which are bounded on both sides by sloping faces are in almost all cases made of equal depth, nor does there seem to be any reason whatever for not adhering to this method.

With these distances thus determined, we return to equations 51 and 54, and from the first of these find the value of v, as was done for equation 63, and substituting for a' the value 40, and for b the quantity 18 feet, we have

$$v=\sqrt{\frac{12960+32000}{98.4}}-18=3.37$$

And, consequently, $b'=b+v=21.37$ feet. To find the value of b', however, it is necessary to use equations 51 and 54, from which by the common method of elimination we may find an expression

$$\theta\, y^3 - v^2 \lambda\, y - 3\, b\, y\, v=0$$

from which by the substitution of the proper values we obtained for a final value of b'', or the thickness of the base of this section, $b''=54.64$ feet, or $v=33.27$ feet. The next step is to find the values of x and y for the third section. As this, and also the last section have both faces sloping, by substituting the value of u given in equation 61, in equations 15 and 16, and reducing and then eliminating, we obtain two expressions for x and y, from which we derive the thickness G F = 100.36, and by a similar process find that for I H to be 152.22 feet.

It is thus apparent, that as there is almost no limit to the number of sections into which a dam may, on this principle, be divided, there are a great number of different forms of profile, each of which, satisfy the conditions of stability, but vary somewhat as to economy. Theoretically the dam whose outer face consists of the greatest number of these sloping faces is the most economical, because in that case its face approaches nearest to the logarithmic curve which bounds the theoretical profile of equal resistance, and it therefore contains very little more masonry than is absolutely necessary to insure safety. In practice, however, such a dam would, in all probability prove much more costly than one consisting of a less number of section, though containing more masonry, because the angle of inclination of the different sections of the outer face changing so frequently would greatly increase the cost of cutting the facing stone. To avoid the mechanical difficulties also likely to arise in such cases, it is sometimes well to depart altogether from this style of profile, and instead of sloping the outer and inner faces, cut them into notches or steps.

THE STEPPED PROFILE.

The stepped profile has been reserved to the last for consideration, because, while it is a natural outgrowth of the preceding modifications, it possesses many merits whose importance cannot be fully appreciated till a comparison is instituted between it and the forms just treated of. In point of simplicity of construction for instance, it would be difficult to find any design of profile that can surpass it. Wherever the faces of the dam are curved as in Fig. 9, or made up of a series of sloping surfaces of various inclination as in Figs. 12, 16 and 17, the dimensions of every facing stone that is set have to be most carefully determined beforehand by the rules of stereography, and this, when the dam is an high one and the number of stones consequently large, is of itself a work of no small difficulty. In the stepped dam however, all this is done away with, as every facing stone, (unless the dam is curved) possesses only a vertical or, if it happens to form the edge of the step, a vertical and horizontal face, and thus requires no pattern for the stone cutter. A further advantage to be derived from it, is, that it enables us to approach much nearer the curved form of profile than we can in any other profile type. Indeed, when well designed it is in reality nothing but the logarithmic curved profile cut into steps or notches, so that should we draw a continuous line through the upper edges of all the steps, or through the lower edges of their vertical faces, this line would form a logarithmic curve.

Here, as in the calculation of the previous profiles, it is quite allowable to assume arbitrarily either the breadth or height of the step and from this one determine the other. Yet it is by far the best plan to assume the vertical height of the step and calculate the breadth. For, it must be apparent, that by this method of procedure, the quantity we calculate is really the abscissa of the curve, which we lay off at regular intervals perpendicularly to the vertical axis of the dam, and in this way we are enabled to preserve very closely the logarithmic profile. The general appearance of the dam is, moreover, much more pleasing when this arrangement is observed than when we assume a constant breadth and calculate the depth, because the breadth of the steps near the summit of the dam is then very narrow and increases gradually as they approach the bottom, and the departure from the curve is thus scarcely perceptible; but when the breadth is everywhere the same and the depth varies, the whole face of the dam has an extremely broken appearance, which is anything but agreeable.

In this profile, as in all the others, the inner face is made vertical for as great a distance as the limit of pressure will allow, and from that point down it is stepped. The outer face is likewise made vertical for a distance which depends in all cases on the thickness across the top, being as a general thing very nearly twice that dimension. In the determination of the following formulæ, the depth of the step has been assumed as the same throughout the entire dam, and the breadth has been taken as the unknown quantity. Fig. 18 then represents a portion of the profile of a dam bound by a curved or sloping face, which we wish to change into a stepped profile. A B D C represents this section, and if H F be taken as the vertical height of the step, then will C H F represent the

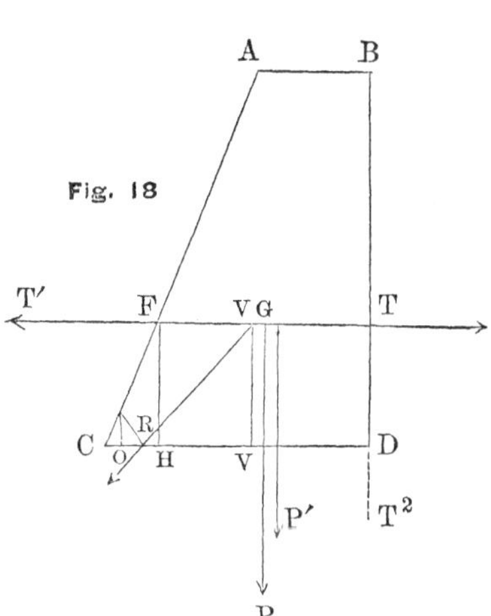

element with which we are especially concerned, and its base C H the quantity we are in search of,—the breadth of the step. The height B D of the section we will denote by h; and the density of the masonry by δ'; and the greatest thickness F T or H D of the *known* element A B T D H F by t; from which three quantities we may obtain an expression for the weight P, of this element, which must of course be accurately known, in-as much as the object of making the step at this point being to lessen the amount of vertical pressure on each superficial unit, the breadth of the step will depend very largely on the weight of that portion of the dam which is above it. The weight which is plainly equal to $\left(\frac{AB + FT}{2}\right) BT\,\delta' + (FH \times HD)\,\delta'$ is expressed by P, while that of the element C H F is equal to $\frac{b\,a\,\delta'}{2}$, in which a is the height of the step F H, and b the breadth C H. The point of application of the thrust of the water is T situated at two-thirds the depth of immersion. T′ and T^2 are the horizontal and vertical components respectively. Then will P represent the direction of the resultants of P and T^2; V V the resultant of P, T^2 and the weight $\frac{a\,b\,\delta'}{2}$ of the element C H F, while the general resultant of all the forces is R. Now, in this case, as in the previous ones, the whole solution of the problem depends on finding the value of C R, or the distance from the outer edge C to the point where the resultant cuts the base, and this we will express as heretofore by the letter u. Then from the figure

$$u = \text{C H} + \text{H V} - \text{R V} \quad . \quad . \quad . \quad 64.$$

in which we know the value of C H$=b$, and require that of H V and R V. But $\frac{\text{R V}}{\text{V V}}$ is equal to the tangent of the angle which the general resultant R makes with the vertical, or calling this angle α then

$$\tan. \alpha = \frac{\text{R V}}{\text{V V}} = \frac{\text{T}'}{\text{P} + \frac{a\, b\, \delta'}{2}}$$

$$\text{R V} = \frac{\text{T}'}{\frac{\text{P} + \frac{a\, b\, \delta'}{2}}{\text{V V}}} = \frac{\text{T}'\, e}{\text{P} + \frac{a\, \delta'\, b}{2}}$$

in which e is to be understood to express the value of V V$=\frac{\text{B D}}{\text{D T}}$. The distance H V may be found from the theorem of moments, by expressing the relation that

$$\text{H V}\times\left(\text{P}+\frac{\delta' a\, b}{2}\right)=\text{M}-\frac{\delta' b^2 a}{6}$$

$$\text{H V}=\frac{\text{M}-\dfrac{\delta' b^2 a}{6}}{\text{P}+\dfrac{\delta' a\, b}{2}}$$

M denoting the moment of P′ with respect of H. As to C H, its value is b, the quantity we are in search of. Replacing these quantities in the equation expressive of the value of u, we have

$$u=b+\frac{\dfrac{\text{M}-\delta' b^2 a}{6}}{\text{P}+\dfrac{\delta' a\, b}{2}}-\frac{\text{T}' e}{\text{P}+\dfrac{a\,\delta' b}{2}}$$

which, reduced to a common denominator, becomes

65.

$$u=\frac{6\, b\,\text{P}+3\, b^2 a\,\delta'+\text{M}-\delta' b^2 a-6\,\text{T}' e}{6\,\text{P}+3\,\delta' b\, a}$$

Having thus obtained an equation for the value of u, the next step is to find by means of it an expression for b the breadth of the step. For this purpose draw from R, the point at which the general resultant of all the acting forces cuts the base, a perpendicular R N to the resultant, and from N a perpendicular to the base C D, thus forming a triangle R N O. Then, since the two triangles R V V and R N O have their bases on the same right line C D, and the side V R of the one perpendicular to the side N R of the other, and the sides V V and N O parallel, the angles at V and N are equal and the triangles are similar. But by the relation existing between the sides of such similar triangles, we have the proportion

$$N\,O : R\,V :: R\,O : V\,V.$$

which gives for N O the equation

$$N\,O = \frac{R\,V \times R\,O}{V\,V} = \frac{T'f}{P + \frac{\delta' a\,b}{2}} \quad . \quad . \quad . \quad 66.$$

in which f is the distance R O. But we have another pair of similar triangles which gives yet another value for N O, which must be deduced and made equal to that just found. These triangles are C O N and C H F, and the proportion derived from the relation of their sides is,

$$N\,O : C\,O :: F\,H : H\,C \text{ or}$$

$$N\,O = \frac{C\,O \times F\,H}{H\,C} = \frac{C\,O \times a}{b} \quad . \quad 67.$$

Equating equations 66 and 67,

$$C\,O \times \frac{a}{b} = f \times \frac{T'}{P + \frac{\delta' a b}{2}}$$

$$C\,O : f :: \frac{T'}{P + \frac{\delta' a b}{2}} : \frac{a}{b}$$

And again, since if four quantities be proportional they will be in proportion by composition and division

$$C\,O + f : f :: \frac{a}{b} + \frac{T'}{P + \frac{\delta' a b}{2}} : \frac{a}{b}$$

and reducing,

$$f=\frac{u\frac{a}{b}}{\frac{T'}{P+\frac{\delta' a b}{2}\times\frac{a}{b}}}=\frac{u\,a\,(\delta' a b+2\,P)}{a(\delta' a b+2\,P)+2\,T' b} \qquad 68.$$

But the condition of stability is (equation 16) expressed by the relation

$$f=\frac{2\left(P+\frac{\delta' a b}{2}\right)}{3\,\delta'\,\lambda} \text{ or } \frac{2\,P+\delta' b\,a}{3\,\delta'\,\lambda} \qquad 69.$$

And equating these values given in equations 68 and 69,

$$\frac{u\,a\,(\delta' a\,b+2\,P)}{a\,(\delta' a\,b+2\,P)+2\,T' b}=\frac{2\,P+\delta' b\,a}{3\,\delta'\,\lambda}$$

Substituting for u its equivalent value as given in equation 65, and dividing both members of the resulting equation by the common factor $2\,P+\delta' b\,a$, there results

$$\delta'\lambda\,a(6\,b\,P'+3\,b^2\delta' a+6\,M-6\,T'e-\delta' b^2 a)$$
$$=a\,(2\,P'+\delta' a\,b+2\,b\,T')\,(2\,P'+\delta'\,a\,b)$$

Solving this with respect to $x\,b^2$, and extracting the root,

$$b = -\frac{P}{\delta'}\,\frac{3\lambda - \dfrac{2T'}{\delta' a} - 2a}{2\lambda a - a^2 - \dfrac{2T'}{\delta'}} + \sqrt{\frac{P^2\left(3\lambda - 2a - \dfrac{2T'}{\delta' a}\right)^2}{\delta'^2\left(a(2\lambda - a - \dfrac{2T}{\delta'}\right)^2} + \frac{\dfrac{P^2}{\delta'^2} + 3\lambda\left(\dfrac{T'e}{\delta'} - \dfrac{M}{\delta'}\right)}{2a\lambda - a^2 - \dfrac{2T'}{\delta'}}}$$

But this is capable of being yet further reduced by dividing through by

$$\frac{3\lambda - \dfrac{2T'}{\delta' a} - 2a}{2\lambda - a - \dfrac{2T'}{\delta'}}$$

to the form

$$b = -\frac{P}{\delta' a} + \sqrt{\frac{P^2}{\delta'^2 a^2} + \frac{\dfrac{4P^2}{\delta'^2} - \lambda\left(6\dfrac{M}{\delta'} + \theta h^3\right)}{3\lambda a - 4a^2}} \qquad 70.$$

which is the expression for the breadth

of the step. As to the meaning of the letters it may once more be stated, that P is the weight in pounds of A B D H F, and δ' the density of the masonry. The *vertical* height (F H), which we determine to give the step, is expressed by a, that of the entire dam from the top to the *base* of the step by h, and the *moment* of the weight P, with respect to the vertical F H forming the *rise* of the step by M ; while by λ, we mean, as in all previous formulæ, the greatest height to which we can raise a vertical wall without the pressure per unit of surface on the base, becoming larger than the limit R′ of pressure ; and by θ, the expression $\frac{\delta}{\delta'}$, or the ratio in which the density of the masonry exceeds that of water. This value of θ, is safely taken at $\frac{1}{2}$. As to the height to be given to the step, this is of course to be assumed at pleasure, but the most pleasing effect is produced when it is taken at six or seven feet, for then, even in dams of one hundred and sixty feet in height, constructed of the heaviest stone, the breadth of the step will rarely at any point be materially greater than the rise. The point on the outer face at which the first step should begin, or in other words the distance A B, in Fig. 19, is deter-

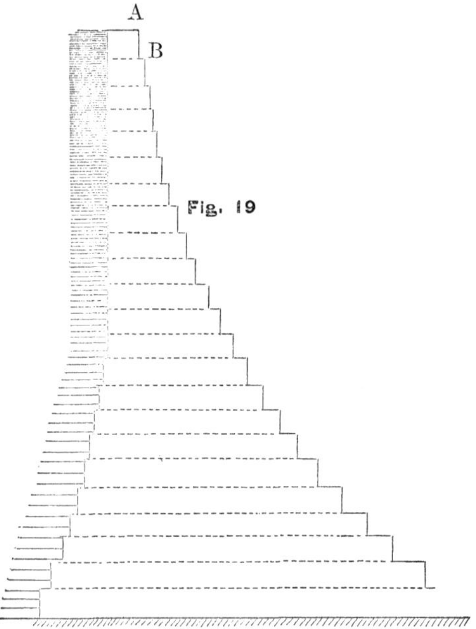

Fig. 19

mined, as in the other instances, by the relation which the breadth on top bears to the height. If the thickness t, across the summit be assumed then

$$a = \sqrt{\frac{3\,t^2}{\theta} + \frac{4\,t^4}{\theta^2\,\lambda^2}} - \frac{2\,t^2}{\theta\,\lambda}$$

but if the height a be assumed the proper thickness is to be had from the equation,

$$t = a\sqrt{\frac{\theta\,\lambda}{3\,\lambda - 4\,a}}$$

When that point on the inner face is reached, at which it becomes necessary to begin stepping, the breadths b and b', of the outer and inner steps respectively, may be had by substituting the value of u, in equations 15 and 16, and from the two resulting equations, finding by elimination two expressions for b and b'. This calculation may, however, be avoided, and considerable expense for cutting facing stone saved, by making the inner face vertical from top to bottom. Indeed the matter of expense for dressing stone is, perhaps, the most serious objection to the stepped profile, as it is necessary to dress both faces of the step.

As regards the use of the formulæ for this form of profile, it is to be borne in mind, that P includes the weight of the water as well as the weight of the masonry, so that in determining the breadth of the fourth step, the weight of the three columns of water resting, one on the first, one on the second and one on the third step, is to be added to the pressure of the masonry. The pressure of the water is readily obtained from equation 1.

The principles that have now been established in connection with the four types of profiles treated of, are all that are required to calculate the parts of any profile that is ever likely to arise in practice. They have, moreover, been determined without regard to the length of the dam, so that the structure will be one of equal resistance, and withstand the thrust of the water solely by its own weight. There is, therefore, no valid reason why a dam constructed with a profile of equal resistance should be curved into the form of an arch, and this holds good, whether it be high or low, whether it obstructs a broad valley or a narrow one. The only thing that can be accomplished by curving a dam, is to relieve it from severe strains, by transmitting as large a part of the thrust to the sides of the valley, but where the profile is such that the dam is everywhere equally strong, and equally capable of resisting by its own weight the severest strain it is ever subjected to, there is surely nothing to be gained by increasing its length in order to transmit this thrust laterally to the sides of the valley. It is true that in deep and narrow valleys, some saving of material may be affected by curving the dam, which being thus relieved from a goodly portion of the thrust, may be diminished in thickness. But in long dams, it is an open question whether the saving thus affected is not more than balanced by the increased length.

One other matter which deserves the

most careful attention, and which indeed unless it is carefully attended to will render the very best profile of no account, it is the binding of the stones, and the character of the inner filling. As to the bond, it is undoubtedly the wisest plan if the dam is to resist a great pressure, to *avoid* laying the stones in horizontal courses wherever such a thing is practicable, and to place *binders* in every possible direction. For assuredly, if it is necessary for the stability of all walls bearing a vertical load, that there should be no continuous joints in the direction of the pressure, it is just as important that a dam should have no *continuous horizontal joints,* because in the case of such structures almost every ounce of thrust they have to resist is horizontal, and thus exactly coincides with the joints. If the dam is *curved,* then this matter of broken horizontal joints is not of such *vital* importance, because no layer can then slide until some one of the stones has been crushed, yet even here it cannot be too rigidly

**** Any book in this Catalogue sent free by mail on receipt of price.*

VALUABLE

SCIENTIFIC BOOKS,

PUBLISHED BY

D. VAN NOSTRAND,

23 MURRAY STREET AND 27 WARREN STREET,

NEW YORK.

FRANCIS. Lowell Hydraulic Experiments, being a selection from Experiments on Hydraulic Motors, on the Flow of Water over Weirs, in Open Canals of Uniform Rectangular Section, and through submerged Orifices and diverging Tubes. Made at Lowell, Massachusetts. By James B. Francis, C. E. 2d edition, revised and enlarged, with many new experiments, and illustrated with twenty-three copperplate engravings. 1 vol. 4to, cloth......................$15 00

ROEBLING (J. A.) Long and Short Span Railway Bridges. By John A. Roebling, C. E. Illustrated with large copperplate engravings of plans and views. Imperial folio, cloth.............................. 25 00

CLARKE (T. C.) Description of the Iron Railway Bridge over the Mississippi River, at Quincy, Illinois. Thomas Curtis Clarke, Chief Engineer. Illustrated with 21 lithographed plans. 1 vol. 4to, cloth.. 7 50

TUNNER (P.) A Treatise on Roll-Turning for the Manufacture of Iron. By Peter Tunner. Translated and adapted by John B. Pearse, of the Pennylvania Steel Works, with numerous engravings wood cuts and folio atlas of plates.................. $10 00

ISHERWOOD (B. F.) Engineering Precedents for Steam Machinery. Arranged in the most practical and useful manner for Engineers. By B. F. Isherwood, Civil Engineer, U. S. Navy. With Illustrations. Two volumes in one. 8vo, cloth........... $2 50

GILLMORE (Gen. Q. A,) Practical Treatise on the Construction of Roads, Streets, and Pavements. By Q. A. Gillmore, Lt.-Col. U. S Corps of Engineers, Brevet Major-Gen. U. S. Army. With 70 illustrations. 12mo, cloth................................ 2 00

—— Report on Strength of the Building Stones in the United States, etc. 8vo, illustrated, cloth..... .... 1 50

CAMPIN on the Construction of Iron Roofs. By Francis Campin. 8vo, with plates, cloth........... 2 00

COLLINS. The Private Book of Useful Alloys and Memoranda for Goldsmiths, Jewellers, &c. By James E. Collins. 18mo, cloth.................... 75

CIPHER AND SECRET LETTER AND TELEGRAPHIC CODE, with Hogg's Improvements. The most perfect secret code ever invented or discovered. Impossible to read without the key. By C. S. Larrabee. 18mo, cloth. 1 00

COLBURN. The Gas Works of London. By Zerah Colburn, C. E. 1 vol 12mo, boards............... 60

CRAIG (B. F.) Weights and Measures. An account of the Decimal System, with Tables of Conversion for Commercial and Scientific Uses. By B. F. Craig, M.D. 1 vol. square 32mo, limp cloth.... 50

NUGENT. Treatise on Optics; or, Light and Sight, theoretically and practically treated; with the application to Fine Art and Industrial Pursuits. By E. Nugent. With one hundred and three illustrations. 12mo, cloth.................................... 2 00

FREE HAND DRAWING. A Gnide to Ornamental Figure and Landscape Drawing. By an Art Student. 18mo, boards,............................. 50

HOWARD. Earthwork Mensuration on the Basis of the Prismoidal Formulae. Containing simple and labor-saving method of obtaining Prismoidal contents directly from End Areas. Illustrated by Examples, and accompanied by Plain Rules for Practical Uses. By Conway R. Howard, C. E., Richmond, Va. Illustrated, 8vo, cloth.............. .... 1 50

GRUNER. The Manufacture of Steel. By M. L. Gruner. Translated from the French, by Lenox Smith, with an appendix on the Bessamer process in the United States, by the translator. Illustrated by Lithographed drawings and wood cuts. 8vo, cloth.. 3 50

AUCHINCLOSS. Link and Valve Motions Simplified. Illustrated with 37 wood-cuts, and 21 lithographic plates, together with a Travel Scale, and numerous useful Tables. By W. S. Auchincloss. 8vo, cloth.. 3 00

VAN BUREN. Investigations of Formulas, for the strength of the Iron parts of Steam Machinery. By J. D. Van Buren, Jr., C. E. Illustrated, 8vo, cloth. 2 00

JOYNSON. Designing and Construction of Machine Gearing. Illustrated, 8vo, cloth... .... 2 00

GILLMORE. Coignet Beton and other Artificial Stone. By Q. A. Gillmore, Major U. S. Corps Engineers. 9 plates, views, &c. 8vo, cloth.................. 2 50

SAELTZER. Treatise on Acoustics in connection with Ventilation. By Alexander Saeltzer, Architect. 12mo, cloth.................................... 2 00

BUTLER (W. F.) Ventilation of Buildings. By W. F. Butler. With illustrations. 18mo, boards......... 50

DICTIONARY of Manufactures, Mining, Machinery, and the Industrial Arts. By George Dodd. 12mo, cloth... 2 00

BOW. A Treatise on Bracing, with its application to Bridges and other Structures of Wood or Iron. By Robert Henry Bow, C. E. 156 illustrations, 8vo, cloth.................................... 1 50

BARBA (J.) The Use of Steel for Constructive Purposes; Method of Working, Applying, and Testing Plates and Brass. With a Preface by A. L. Holley, C. E. 12mo, cloth.............................. 1 50

GILLMORE (Gen. Q. A.) Treatise on Limes, Hydraulic Cements, and Mortars. Papers on Practical Engineering, U. S. Engineer Department, No. 9, containing Reports of numerous Experiments conducted in New York City, during the years 1858 to 1861, inclusive. By Q. A. Gillmore, Bvt. Maj-Gen., U. S. A., Major, Corps of Engineers. With numerous illustrations. 1 vol, 8vo, cloth.............. $4 00

HARRISON. The Mechanic's Tool Book, with Practical Rules and Suggestions for Use of Machinists, Iron Workers, and others. By W. B. Harrison, associate editor of the "American Artisan." Illustrated with 44 engravings. 12mo, cloth............ 1 50

HENRICI (Olaus). Skeleton Structures, especially in their application to the Building of Steel and Iron Bridges. By Olaus Henrici. With folding plates and diagrams. 1 vol. 8vo, cloth................... 1 50

HEWSON (Wm.) Principles and Practice of Embanking Lands from River Floods, as applied to the Levees of the Mississippi. By William Hewson, Civil Engineer. 1 vol. 8vo, cloth....................... 2 00

HOLLEY (A. L.) Railway Practice. American and European Railway Practice, in the economical Generation of Steam, including the Materials and Construction of Coal-burning Boilers, Combustion, the Variable Blast, Vaporization, Circulation, Superheating, Supplying and Heating Feed-water, etc., and the Adaptation of Wood and Coke-burning Engines to Coal-burning; and in Permanent Way, including Road-bed, Sleepers, Rails, Joint-fastenings, Street Railways, etc., etc. By Alexander L. Holley, B. P. With 77 lithographed plates. 1 vol. folio, cloth.... 12 00

KING (W. H.) Lessons and Practical Notes on Steam, the Steam Engine, Propellers, etc., etc., for Young Marine Engineers, Students, and others. By the late W. H. King, U. S. Navy. Revised by Chief Engineer J. W. King, U. S. Navy. Nineteenth edition, enlarged. 8vo, cloth....................... 2 00

MINIFIE (Wm.) Mechanical Drawing. A Text-Book of Geometrical Drawing for the use of Mechanics and Schools, in which the Definitions and Rules of Geometry are familiarly explained; the Practical Problems are arranged, from the most simple to the more complex, and in their description technicalities are avoided as much as possible. With illustrations for Drawing Plans, Sections, and Elevations of Railways and Machinery; an Introduction to Isometrical Drawing, and an Essay on Linear Perspective and Shadows. Illustrated with over 200 diagrams engraved on steel. By Wm. Minifie, Architect. Ninth edition. With an Appendix on the Theory and Application of Colors. 1 vol. 8vo, cloth.............. $4 00

"It is the best work on Drawing that we have ever seen, and is especially a text-book of Geometrical Drawing for the use of Mechanics and Schools. No young Mechanic, such as a Machinists, Engineer, Cabinet-maker, Millwright, or Carpenter, should be without it."—*Scientific American.*

—— Geometrical Drawing. Abridged from the octavo edition, for the use of Schools. Illustrated with 48 steel plates. Fifth edition. 1 vol. 12mo, cloth.... 2 00

STILLMAN (Paul.) Steam Engine Indicator, and the Improved Manometer Steam and Vacuum Gauges—their Utility and Application. By Paul Stillman. New edition. 1 vol. 12mo, flexible cloth........... 1 00

SWEET (S. H.) Special Report on Coal; showing its Distribution, Classification, and cost delivered over different routes to various points in the State of New York, and the principal cities on the Atlantic Coast. By S. H. Sweet. With maps, 1 vol. 8vo, cloth..... 3 00

WALKER (W. H.) Screw Propulsion. Notes on Screw Propulsion: its Rise and History. By Capt. W. H. Walker, U. S. Navy. 1 vol. 8vo, cloth..... 75

WARD (J. H.) Steam for the Million. A popular Treatise on Steam and its Application to the Useful Arts, especially to Navigation. By J. H. Ward, Commander U. S. Navy. New and revised edition. 1 vol. 8vo, cloth................................. 1 00

WIESBACH (Julius). A Manual of Theoretical Mechanics. By Julius Weisbach, Ph. D. Translated from the fourth augmented and improved German edition, with an Introduction to the Calculus, by Eckley B. Coxe, A. M., Mining Engineer. 1,100 pages, and 902 wood-cut illustrations. 8vo, cloth 10 00

DIEDRICH. The Theory of Strains, a Compendium for the calculation and construction of Bridges, Roofs, and Cranes, with the application of Trigonometrical Notes, containing the most comprehensive information in regard to the Resulting strains for a permanent Load, as also for a combined (Permanent and Rolling) Load. In two sections, adadted to the requirements of the present time. By John Diedrich, C. E. Illustrated by numerous plates and diagrams. 8vo, cloth.. 5 00

WILLIAMSON (R. S.) On the use of the Barometer on Surveys and Reconnoissances. Part I. Meteorology in its Connection with Hypsometry. Part II. Barometric Hypsometry. By R. S. Wiliamson, Bvt Lieut.-Col. U. S. A., Major Corps of Engineers. With Illustrative Tables and Engravings. Paper No. 15, Professional Papers, Corps of Engineers. 1 vol. 4to, cloth.................................. 15 00

POOK (S. M.) Method of Comparing the Lines and Draughting Vessels Propelled by Sail or Steam. Including a chapter on Laying off on the Mould-Loft Floor. By Samuel M. Pook, Naval Constructor. 1 vol. 8vo, with illustrations, cloth............ 5 00

ALEXANDER (J. H.) Universal Dictionary of Weights and Measures, Ancient and Modern, reduced to the standards of the United States of America. By J. H. Alexander. New edition, enlarged. 1 vol. 8vo, cloth.................................. 3 50

WANKLYN. A Practical Treatise on the Examination of Milk, and its Derivatives, Cream, Butter and Cheese. By J. Alfred Wanklyn, M. R. C. S., 12mo, cloth.. 1 00

RICHARDS' INDICATOR. A Treatise on the Richards Steam Engine Indicator, with an Appendix by F. W. Bacon, M. E. 18mo, flexible, cloth.......... 1 00

PORTER (C. T.) A Treatise on the Richards Steam Engine Indicator, and the Development and Application of Force in the Steam Engine. By Charles T. Porter. Third edition, revised and enlarged. 8vo, illustrated, cloth.................................. 3 50

POPE. Modern Practice of the Electric Telegraph. A Hand Book for Electricians and operators. By Frank L. Pope. Ninth edition, revised and enlarged, and fully illustrated. 8vo, cloth........................ $2 00

"There is no other work of this kind in the English language that contains in so small a compass so much practical information in the application of galvanic electricity to telegraphy. It should be in the hands of every one interested in telegraphy, or the use of Batteries for other purposes."

EASSIE (P. B.) Wood and its Uses. A Hand-Book for the use of Contractors, Builders, Architects, Engineers, and Timber Merchants. By P. B. Eassie. Upwards of 250 illustrations. 8vo, cloth........... 1 50

SABINE. History and Progress of the Electric Telegraph, with descriptions of some of the apparatus. By Robert Sabine, C. E. Second edition, with additions, 12mo, cloth...... .... 1 25

BLAKE. Ceramic Art. A Report on Pottery, Porcelain, Tiles, Terra Cotta and Brick. By W. P. Blake, U. S. Commissioner, Vienna Exhibition, 1873. 8vo, cloth.. 2 00

BENET. Electro-Ballistic Machines, and the Schultz Chronoscope. By Lieut.-Col. S. V. Benet, Captain of Ordnance, U. S. Army. Illustrated, second edition, 4to, cloth................................ 3 00

MICHAELIS. The Le Boulenge Chronograph, with three Lithograph folding plates of illustrations. By Brevet Captain O. E. Michaelis, First Lieutenant Ordnance Corps, U. S. Army, 4to, cloth.......... 3 00

ENGINEERING FACTS AND FIGURES An Annual Register of Progress in Mechanical Engineering and Construction, for the years 1863, 64, 65, 66 67, 68. Fully illustrated, 6 vols. 18mo, cloth, $2.50 per vol., each volume sold separately..............

HAMILTON. Useful Information for Railway Men. Compiled by W. G. Hamilton, Engineer. Sixth edition, revised and enlarged, 562 pages Pocket form. Morocco, gilt.................................... 2 00

STUART (B.) How to Become a Successful Engineer. Being Hints to Youths intending to adopt the Profession. Sixth edition. 12mo, boards............ 50

STUART. The Civil and Military Engineers of America. By Gen. C. B. Stuart. With 9 finely executed portraits of eminent engineers, and illustrated by engravings of some of the most important works constructed in America. 8vo, cloth.................. $5 00

STONEY. The Theory of Strains in Girders and similar structures, with observations on the application of Theory to Practice, and Tables of Strength and other properties of Materials. By Bindon B. Stoney, B. A. New and revised edition, enlarged, with numerous engravings on wood, by Oldham. Royal 8vo, 664 pages. Complete in one volume. 8vo, cloth....... 12 50

SHREVE. A Treatise on the Strength of Bridges and Roofs. Comprising the determination of Algebraic formulas for strains in Horizontal, Inclined or Rafter, Triangular, Bowstring, Lenticular and other Trusses, from fixed and moving loads, with practical applications and examples, for the use of Students and Engineers. By Samuel H. Shreve, A. M., Civil Engineer. 87 wood-cut illustrations. 2d edition. 8vo, cloth... 5 00

MERRILL. Iron Truss Bridges for Railroads. The method of calculating strains in Trusses, with a careful comparison of the most prominent Trusses, in reference to economy in combination, etc., etc. By Brevet Col. William E. Merrill, U S. A., Major Corps of Engineers, with nine lithographed plates of Illustrations. 4to, cloth.......................... 5 00

WHIPPLE. An Elementary and Practical Treatise on Bridge Building. An enlarged and improved edition of the author's original work. By S. Whipple, C. E., inventor of the Whipple Bridges, &c. Illustrated 8vo, cloth...................................... 4 00

THE KANSAS CITY BRIDGE. With an account of the Regimen of the Missouri River, and a description of the methods used for Founding in that River. By O. Chanute, Chief Engineer, and George Morrison, Assistant Engineer. Illustrated with five lithographic views and twelve plates of plans. 4to, cloth, 6 00

DUBOIS (A. J.) The New Method of Graphical Statics. By A. J. Dubois, C. E., Ph. D. With 60 illustrations. 8vo, cloth 2 00

MAC CORD. A Practical Treatise on the Slide Valve by Eccentrics, examining by methods the action of the Eccentric upon the Slide Valve, and explaining the Practical processes of laying out the movements, adapting the valve for its various duties in the steam engine. For the use of Engineers, Draughtsmen, Machinists, and Students of Valve Motions in general. By C. W. Mac Cord, A. M., Professor of Mechanical Drawing, Stevens' Institute of Technology, Hoboken, N. J. Illustrated by 8 full page copper-plates. 4to, cloth.................................. $4 00

KIRKWOOD. Report on the Filtration of River Waters, for the supply of cities, as practised in Europe, made to the Board of Water Commissioners of the City of St. Louis. By James P. Kirkwood. Illustrated by 30 double plate engravings. 4to, cloth, 15 00

PLATTNER. Manual of Qualitative and Quantitative Analysis with the Blow Pipe. From the last German edition, revised and enlarged. By Prof. Th. Richter, of the Royal Saxon Mining Academy. Translated by Prof. H. B. Cornwall, Assistant in the Columbia School of Mines, New York assisted by John H. Caswell. Illustrated with 87 wood cuts, and one lithographic plate. Third edition, revised, 560 pages, 8vo, cloth.. 7 50

PLYMPTON. The Blow Pipe. A Guide to its Use in the Determination of Salts and Minerals. Compiled from various sources, by George W. Plympton, C. E. A. M., Professor of Physical Science in the Polytechnic Institute, Brooklyn, New York, 12mo, cloth.. 1 50

PYNCHON. Introduction to Chemical Physics, designed for the use of Academies, Colleges and High Schools. Illustrated with numerous engravings, and containing copious experiments with directions for preparing them. By Thomas Ruggles Pynchon, M. A., Professor of Chemistry and the Natural Sciences, Trinity College, Hartford New edition, revised and enlarged, and illustrated by 269 illustrations on wood. Crown, 8vo. cloth...................... 3 00

ELIOT AND STORER. A compendious Manual of Qualitative Chemical Analysis. By Charles W. Eliot and Frank H. Storer. Revised with the Co-operation of the authors. By William R. Nichols, Professor of Chemistry in the Massachusetts Institute of Technology Illustrated, 12mo, cloth....... $1 50

RAMMELSBERG. Guide to a course of Quantitative Chemical Analysis, especially of Minerals and Furnace Products. Illustrated by Examples By C. F. Rammelsberg. Translated by J. Towler, M. D. 8vo, cloth.. 2 25

DOUGLASS and PRESCOTT. Qualitative Chemical Analysis. A Guide in the Practical Study of Chemistry, and in the Work of Analysis. By S. H. Douglass and A. B. Prescott, of the University of Michigan. New edition. 8vo. *In press.*

JACOB. On the Designing and Construction of Storage Reservoirs, with Tables and Wood Cuts representing Sections, &c., 18mo, boards.......................... 50

WATT'S Dictionary of Chemistry. New and Revised edition complete in 6 vols 8vo cloth, $62.00 Supplementary volume sold separately. Price, cloth... 9 00

RANDALL. Quartz Operators Hand-Book. By P. M. Randall. New edition, revised and enlarged, fully illustrated. 12mo, cloth 2 00

SILVERSMITH. A Practical Hand-Book for Miners, Metallurgists, and Assayers, comprising the most recent improvements in the disintegration amalgamation, smelting, and parting of the precious ores, with a comprehensive Digest of the Mining Laws Greatly augmented, revised and corrected. By Julius Silversmith. Fourth edition. Profusely illustrated. 12mo, cloth.. 3 00

THE USEFUL METALS AND THEIR ALLOYS, including Mining Ventilation, Mining Jurisprudence, and Metallurgic Chemistry employed in the conversion of Iron, Copper, Tin, Zinc, Antimony and Lead ores, with their applications to the Industrial Arts. By Scoffren, Truan, Clay, Oxland, Fairbairn, and others. Fifth edition, half calf...................... 3 75

JOYNSON. The Metals used in construction, Iron, Steel, Bessemer Metal, etc., etc. By F. H. Joynson. Illustrated, 12mo, cloth.......................... $0 75

VON COTTA. Treatise on Ore Deposits. By Bernhard Von Cotta, Professor of Geology in the Royal School of Mines, Freidberg, Saxony. Translated from the second German edition, by Frederick Prime, Jr., Mining Engineer, and revised by the author, with numerous illustrations. 8vo, cloth....... 4 00

GREENE. Graphical Method for the Analysis of Bridge Trusses, extended to continuous Girders and Draw Spans. By C. E. Greene, A. M., Prof. of Civil Engineering, University of Michigan. Illustrated by 3 folding plates, 8vo, cloth.......................... 2 00

BELL. Chemical Phenomena of Iron Smelting. An experimental and practical examination of the circumstances which determine the capacity of the Blast Furnace, The Temperature of the air, and the proper condition of the Materials to be operated upon. By I. Lowthian Bell. 8vo, cloth..... 6 00

ROGERS. The Geology of Pennsylvania. A Government survey, with a general view of the Geology of the United States, Essays on the Coal Formation and its Fossils, and a description of the Coal Fields of North America and Great Britain. By Henry Darwin Rogers, late State Geologist of Pennsylvania, Splendidly illustrated with Plates and Engravings in the text. 3 vols, 4to, cloth with Portfolio of Maps. 30 00

BURGH. Modern Marine Engineering, applied to Paddle and Screw Propulsion. Consisting of 36 colored plates, 259 Practical Wood Cut Illustrations, and 403 pages of descriptive matter, the whole being an exposition of the present practice of James Watt & Co., J & G Rennie, R. Napier & Sons, and other celebrated firms, by N. P. Burgh, Engineer, thick 4to, vol., cloth, $25.00; half mor.. ..... 30 00

CHURCH. Notes of a Metallurgical Journey in Europe. By J. A. Church, Engineer of Mines, 8vo, cloth..... 2 00

BOURNE. Treatise on the Steam Engine in its various applications to Mines, Mills, Steam Navigation, Railways, and Agriculture, with the theoretical investigations respecting the Motive Power of Heat, and the proper proportions of steam engines. Elaborate tables of the right dimensions of every part, and Practical Instructions for the manufacture and management of every species of Engine in actual use. By John Bourne, being the ninth edition of "A Treatise on the Steam Engine," by the "Artizan Club." Illustrated by 38 plates and 546 wood cuts. 4to, cloth..$15 00

STUART. The Naval Dry Docks of the United States. By Charles B. Stuart late Engineer-in-Chief of the U. S. Navy. Illustrated with 24 engravings on steel. Fourth edition, cloth.................... 6 00

ATKINSON. Practical Treatises on the Gases met with in Coal Mines. 18mo, boards................ 50

FOSTER. Submarine Blasting in Boston Harbor, Massachusetts. Removal of Tower and Corwin Rocks. By J. G. Foster, Lieut.-Col. of Engineers, U. S. Army. Illustrated with seven plates, 4to, cloth.. 3 50

BARNES Submarine Warfare, offensive and defensive, including a discussion of the offensive Torpedo System, its effects upon Iron Clad Ship Systems and influence upon future naval wars. By Lieut.-Commander J. S. Barnes, U. S. N., with twenty lithographic plates and many wood cuts. 8vo, cloth..... 5 00

HOLLEY. A Treatise on Ordnance and Armor, embracing descriptions, discussions, and professional opinions concerning the materials, fabrication, requirements, capabilities, and endurance of European and American Guns, for Naval, Sea Coast, and Iron Clad Warfare, and their Rifling, Projectiles, and Breech-Loading; also, results of experiments against armor, from official records, with an appendix referring to Gun Cotton, Hooped Guns, etc., etc. By Alexander L. Holley, B. P., 948 pages, 493 engravings, and 147 Tables of Results, etc., 8vo, half roan. 10 00

www.ingramcontent.com/pod-product-compliance
Lightning Source LLC
LaVergne TN
LVHW021411110826
845150LV00007B/1874

* 9 7 8 1 4 2 5 5 1 0 7 8 7 *